The Apatani Way of Life

This book celebrates the heritage of the distinctive Apatani community of the north-eastern Himalayan state of Arunachal Pradesh in India. It explores the fascinating indigenous knowledge of field and forest and a uniquely sustainable and enduring way of life that continues to evolve in the modern context. The book tells the story of how a material culture was shaped around bamboo and cane resources and nurtured by a strong community spirit and spirituality that transcended the human world and maintained an unbroken ethos of conservation through time. It highlights the eco-sensitive lifestyle of this unique community and presents an in-depth analysis of the Apatani tradition of the exemplary use of natural resources.

Through this engrossing detailed study, the author observes how bamboo houses are built in three days, fish cultivated in a rice field and a single river used for millennia to feed an entire community. She highlights the triumph of the human spirit in engineering a cultural landscape out of a swamp, and how peaceful co-existence with nature can withstand the trials of time.

Part autobiographical and powerfully personal, this book is a primer on sustainable living as practice. It will be of interest to researchers and students of tribal and Himalayan vernacular architecture, traditional bamboo-cane craft, urban ecology and geography, cultural studies, and sustainability. It will also attract general readership while being academically useful for anthropologists, sociologists, botanists, ecologists and environmentalists.

Ritu-Ngapnon Varuni better known as Ritu Varuni is a trained architect and designer, who wears many hats that include natural architecture, craft and furniture design, writing and poetry to creative thinking and ethical learning, and natural Himalayan farming. After completing her research thesis in Nagaland in the nineties, she was based in Arunachal Pradesh and Assam for six years, working on researching and developing the use of bamboo in housing and craft along with other grassroots work. Her design work under her studio name of 'E'thaan', has centred around traditional Indian craft design and skill development for most of her working life and has been exhibited in India and internationally. She has written many articles for books and journals.

Currently based in Himachal Pradesh, Varuni is the founder-director of the Himalayan Rilung Foundation, which is focused on developing a hands-on, skill-based experiential learning programme centred around the Himalayan environment called 'Mountainwind'.

This book is dedicated to the next generation of Apatani youth with hope that they will continue to nurture the Earth as their ancestors did.........

The Apatani Way of Life

Shaping a Culture Through Bamboo, Cane and Land Use

Ritu-Ngapnon Varuni

Routledge
Taylor & Francis Group
LONDON AND NEW YORK

First published 2023
by Routledge
4 Park Square, Milton Park, Abingdon, Oxon OX14 4RN

and by Routledge
605 Third Avenue, New York, NY 10158

Routledge is an imprint of the Taylor & Francis Group, an informa business

British Library Cataloguing-in-Publication Data
A catalogue record for this book is available from the British Library

ISBN: 978-1-032-24713-7 (hbk)
ISBN: 978-1-032-30520-2 (pbk)
ISBN: 978-1-003-30553-8 (ebk)

DOI: 10.4324/9781003305538

Typeset in Times New Roman
by Newgen Publishing UK

Contents

Figures

Tables

Foreword

When Ritu asked me to write the Foreword for this book, I said, sure; because that's what you say when Ritu asks you do something for her. I sat down to write and quickly realised that I had no idea how to write a foreword for a book. Actually, when picking up a new book, I tend to skip the foreword and go straight to the first chapter. You may decide to do that as well. But just in case you don't, here goes.

Ritu and I were fellow travellers in Arunachal Pradesh, although she lived there and I did not. Arunachal is in north-eastern India, the name meaning "land of the rising sun", wholly appropriate for its far-eastern location. We came together and first met there; she, an architect, studying the vernacular architecture, Apatani land-use and the traditional uses of bamboo and cane, and I as a forester, studying the Blue Pine in the area. Along with a local forest government officer, Pekyom Ringu, we travelled into the homeland of the Apatani people and then trekked to Talle Valley. Talle Valley is a sacred forest within the Apatani Valley of Arunachal Pradesh. It is now designated as an official wildlife reserve, but in ancient historical time, it was a way stop on the supposed migration route of the Apatani from the Tibetan border to their present homeland. They brought along with them on that migration, their bamboo-*bije* and the Blue Pine.

There in the Eastern Himalayas, miles from nowhere, in a crude herder's shelter, we heard Ritu sing:

Imagine all the people,
Living for today….
Imagine all the people,
Living Life in Peace…
Imagine all the people sharing all the world.
You may say I'm a dreamer, but I'm not the only one.
I hope someday you'll join us.
And the world will live as one.

(From *Imagine*, by John Lennon)

That is Ritu Varuni. That is what her life is and continues to be through many environmental, spiritual and social initiatives, culminating in her present work – the 'Mountainwind' creative learning programme and creation of an ecologically grounded global family in her home state of Himachal Pradesh, India. That is what you are going to get out of this book, *The Apatani Way of Life*.

The first chapter takes you into the very different way of life of the Apatani people. Their values, sense of community, and their spirituality. From there, in Chapter 2 the journey looks at the ecologically sustainable landscape this culture created. A patchwork of rice lands, natural forests and private plantations of pine and bamboo that supply the daily needs of the community. Chapters 3 and 4 focus on bamboo and cane, their diversity, qualities and uses related to the Apatani Valley. Think of bamboo as the plastic of the Apatani world. Like plastic, it serves most every daily need of today's human world except, fortunately, it's not like plastic at all. It's not made from petroleum, it's been sustainably grown for eons, and it's easily biodegradable. The book ends in Chapter 5, looking at how this Apatani world is being impacted by the global forces around it. Ritu ends on a note of hope for the Apatani world, and maybe even hope for our wider one. All you need to do is *Imagine* a society that adopted just some of the cultural, ecological and spiritual lessons the Apatani have to teach us.

To paraphrase, "*They may say you're a dreamer, but you're not the only one. I hope someday you'll join us, and the world will live as one*".

Enjoy.

Dr Michael Rechlin, West Virginia, USA

Introduction

Can we be human in the absence of nature
Give back the wilderness, take away the city
Come back evenings when herds returned suffused in evening light
Serene hymns were sung, paddy accepted as alms and bark clothes worn.

(From Chaitali, 1896,
Translated by Fakrul Alam and written by
Rabindranath Tagore, Sabhyatar-Prati.)

At the outset, I would like to begin with the redefinition of the words 'tribe' and 'civilisation', if I may. The usage of these interchangeably in the book is deliberate. To my mind, these words, 'tribe' and 'civilisation', can actually be very misleading and have been used to compare cultures as inferior and superior in a sense. In fact, they have also been used either to denigrate, eulogise or romanticise communities. They can be divisive for human beings and have been mostly used as such. Every human being has tribal roots, which to me means a deep connect with the forest and earth to which we all humans inherently belong. Therefore, I like to use the word 'tribe' with an implication of roots rather than in actual communal terms; for it simply indicates our common undeniable origin. So, while many people have lost this connect (sadly to my mind), some of us still retain or even reconnect with these roots as the case may be and thus could be called 'tribal', even individually. The word *Tribe* in the Cambridge dictionary is strangely defined as "*a group of people, often of related families, who live together, sharing the same language, culture, and history, especially those who do not live in towns or cities*". This definition applies to the word 'community', too, mostly except for the underlined urban resident bit! 'Civilisation' is similarly dictionary defined as "*a highly developed culture, including its social organisation, government, laws, and arts, or the culture of a social group or country at a particular time*". How and who designates this 'highly developed' and why? I would like to redefine 'civilisation' as a continuously evolving cultural ethos, which is associated with a community that has settled in a specific geographical region for a length of time and not to be too fussed about it. In all of recorded history, human beings have been behaviourally both savage and civilised at the same time and continue much in the same vein.

Labels have never really worked for me. All given labels, are often limiting, a tight fit or simply misconstrued, and seem to serve a convenient differentiation rather than a real purpose. In a given context, however, I guess they seek to define the character of a person or a community, as I attempt to do in this book.

We lose a lot when we lose our tribal or forest roots. We lose a large part of what makes us human. We lose our connect with nature and the hands-on sustainable lifestyle that goes with it. We uproot our hand and land-based life skills from the core of our lives and push them aside as belonging to a past era. We disregard their value in our lives and the learning mode and way of life that goes with them. Technology has now overshadowed every aspect of our lives in an almost-dominant governing mode that has enslaved minds and habits therein. This has complicated and unceasingly continues to complicate our lives with a highly debatable, so-called 'development', that stifles our ancestral relationship with nature. This is not to entirely deny technology its place in our lives but just to position it in a way that does not overwhelm or destroy our nature-connect and humaneness. We have created a human-made artificial world that causes us to increasingly wonder how we managed to go so wrong. So complex is this 'technilisation', that we have lost the basic simplicity that cradled true humility and understood our insignificance in this basically unknown and mystical universe. We disconnect ourselves from an unavoidably uncertain and impermanent reality and create our own delusionary world in an already truly illusory existence. This I feel deeply is where we went wrong. This is why we find ourselves in dire straits with a greying sky, parched soil and contaminated ocean. To learn what needs to be learnt, to keep what needs to be nurtured, to respect centuries of experience and connect with the natural environment is something we rarely do well nowadays. This was not always the case.

This book is born of the study and analysis of the evergreen legacy (in both senses of the word) of the Apatani civilisation. It turns the spotlight on the eco-sensitive lifestyle of this unique community that occupies the Ziro Valley in the Lower Subansiri district of the north-easternmost Indian state of Arunachal Pradesh. This large, diverse, and beautiful state of India, Arunachal Pradesh was my home and base throughout the nineties. During this period, I lived in the Apatani Valley for five months through 1998–1999 to research and document the indigenous knowledge systems of the Apatani community. My field research at the time focused on the study and utilisation of natural resources with an emphasis on bamboo and cane and traditional knowledge pertaining to the same. Before this I had, in 1992–1993, under the aegis of an NGO, TARU, for development, also done a field research study analysis of the then-existing Apatani house design and construction style. This was part of a team project to design and formulate a ten-year housing action plan for the entire state, a plan that our NGO team at the time had undertaken for the Planning Commission (Government of India).

The Apatani community is living proof of how human beings and nature can always cohabit harmoniously. They remain an outstanding and practical statement of conservation that has withstood a fast-changing and endangered world. The learning that comes from such an inspirational example is indisputable. As an intensive analytical study of the Apatani cultural landscape, with a strong focus on the bamboo and cane resource base and its utilisation and management, the book provides original content based on my unpublished primary field research done in the nineties. So, I write of the Apatani people, not from the outside, but as someone who came close and, if only for a while, stayed to study, document, and learn from what they had to offer to the world. This is a narrative of that personal journey and thus has a biographical undertone.

The Apatani understood how to live with land and forest, without disturbing their intricate natural balance and mystical working. Their traditional knowledge related to the sustainable use of limited local natural resources is underlined by admirably efficient natural agricultural practices and an ingenious land-use system. Despite overpowering external influences, the Apatani culture remains consistent in many of its facets, and it is this uniquely determined character that has inspired the inclusion of its cultural landscape as a tentative entry on the UNESCO World Heritage List in 2017. This book gives them the attention they most certainly deserve.

This personalised document brings to life a detailed and intimate exploration of a uniquely enduring, nature-oriented way of life. It does not mummify a cultural ethos and deny inevitable change, but celebrates the spirit of an ancient value system and tribal lifestyle that had a deep connect with the natural world and its fragile ecology. The earth is our only known home, and we are part of its creatures of the forest; not of the arrogant artificial ancestry we create and the unnecessary divisiveness we foster between ourselves. The more we distance ourselves from both land and forest, the more we will destroy them and eventually in the end ourselves. Why not treasure and relearn from the cultures that still belong to the land and forest. Why not cherish and appreciate once again the value of what gives us life.

Let us find our way back to our forest roots. Let us go back to the basics.

Ritu-Ngapnon Varuni

Acknowledgements

With grateful thanks for their help and guidance without which this study and learning experience would not have been possible.

In the Ziro Valley …

Dr Tage Kanno, Hapoli, for his many insights into change and his invaluable inputs that have enriched this book;

My companions and guides in the forests of Ziro Valley;

Mr Nending Rido, Hija Village, Ms Dani Jenny, Hapoli and Mr Buru Loder, Modang Tage Village;

Ms Taliang Shanti, Mr Michi, Michi Bamin Village;

Mr Punyo Tatung, Hong Village, Mr Kago Kamar, Hong Village;

Mr Hage Gayu, Hari Village; Mr Hage Duliang, Hari Village;

Mr Modang Tara, Modang Tage Village; Mr Dandhin Donny, Modang Tage Village;

Mr Tiling Doley, Ziro; Dr Dani Duri and Mr Dani Loma, Hapoli.

And elsewhere …

G.B Pant Institute of Himalayan Development and Environment (N.E Unit) for funding the major part of my field research project that enabled a large portion of this book;

Dr Haridasan, SFRI, Itanagar for sharing his immense knowledge of Arunachal's flora and guiding my field-based research;

Mr Nani Sha, Itanagar;

Mr P. Ringu, DFO, Itanagar;

Professor Dr. Michael A. Rechlin, University of Chicago, for pointing out the trees from the forest in Talle Valley, the many unforgettable interactions together in Ziro and all the support he has given at every stage of publication of this book;

Professor Dr Savyasaachi, Jamia Milia Islamia University, New Delhi, for giving impetus and guidance in making this publication possible.

1 The Cultural Landscape

Figure 1.0 Bird's-eye view of Ziro Valley, 2021.
Source: Dr Tage Kanno, Hapoli.

In both its diverse geography and content, Arunachal Pradesh, literally meaning the 'Land of the Dawn-Lit Mountains', remains unmatched in the Indian sub-continent as a state with the greatest Himalayan biodiversity. It extends across the three parallel Himalayan ranges from the high altitude Greater Himalayan tract, with snow-laden coniferous forests, to the lush semi-evergreen rain forests that occupy the lower southern boundaries of the state and merge into the Patkai Hills. Its forested area accounts for one-third of the habitat area within the Himalayan biodiversity hot-spot and 81 per cent of its area is under forest cover (refer 2017 Government of Arunachal Pradesh portal www.arunachalpradeshgov.in). A huge variety of vegetation – from pine, fir, deodar and oak to bamboo, cane and *hollock*, *hollong* and teak sustains its myriad wildlife. Leopard, bear, fox, bison, gaur, giant flying

DOI: 10.4324/9781003305538-1

squirrels, tiger and elephant are some of the huge gamut of fauna that can be found here along with a host of colourful bird life that is equally extensive. The *mithun*, or Indian bison (*Bos fontalis*), is the state animal and is intertwined with the socio-cultural ethos of many of the ethnic communities of the state; while the hornbill (*Buceros bicornis*) is the state bird.

The lower reaches have a sub-tropical hot and humid climate with mild winters. The upper altitudes have either an alpine or a sub-Himalayan climate with severely cold winters and pleasantly warm summers. Heavy rainfall (about 5,000 mm annually) almost six months of the year from May to October is experienced in most areas of the state. With the lowest density of population in India at 17 persons per sq km or a total population figure of 13.8 million (2011 Census of India), Arunachal has a large concentration of settlements in the river valleys of the sub-Himalayan and Lower Himalayan tracts. The upper reaches of the state are largely inhospitable and have inaccessible terrain. The projected population of the state (Government of Arunachal Pradesh portal) in 2022 stands at 17.12 lakhs or 17.12 million people.

Journeying to the Ziro Valley

Let me begin this book in a fairy tale fashion. It should have the quality that dreams are made of, for the research proposal that provided most of the material for this book was supposed to be the start and part of a dream. '*Khwaab*' (meaning 'dream' in Urdu and Hindi) was the name I gave to the non-profit organisation that I tried to set up in Itanagar in India in the nineties. Itanagar is the state capital of India's most fascinatingly diverse eastern frontier of Arunachal Pradesh, which is the first state in the country to welcome the morning sun, and was my home through the nineties. From the snow–covered, high-altitude mountains to the low-lying tropical forests and flat river valleys, its geographical diversity is matched by the myriad cultures it has created in its embrace. This huge land mass spread over eighty thousand square kilometres (83,743 sq km) is one of India's largest states and home to 25 major tribal communities with over a hundred sub-tribes speaking almost as many distinct languages. It has three international borders with Tibet (now under China) along its northern limit, Myanmar on its eastern periphery and Bhutan touching its western boundary. It cradles the Indian state of Assam on its long southern side and a small area of its south eastern corner also touches the neighbouring Indian state of Nagaland (see Location Map Figure 1.1). With turbulent mountain rivers running north–south and bifurcating its thickly forested expanse into three distinct zones, this stunningly beautiful eastern Himalayan state was a hidden and mostly inaccessible place until the late eighteenth century, when the British referred to it as NEFA (North-east Frontier Agency). An erstwhile Union territory governed centrally for

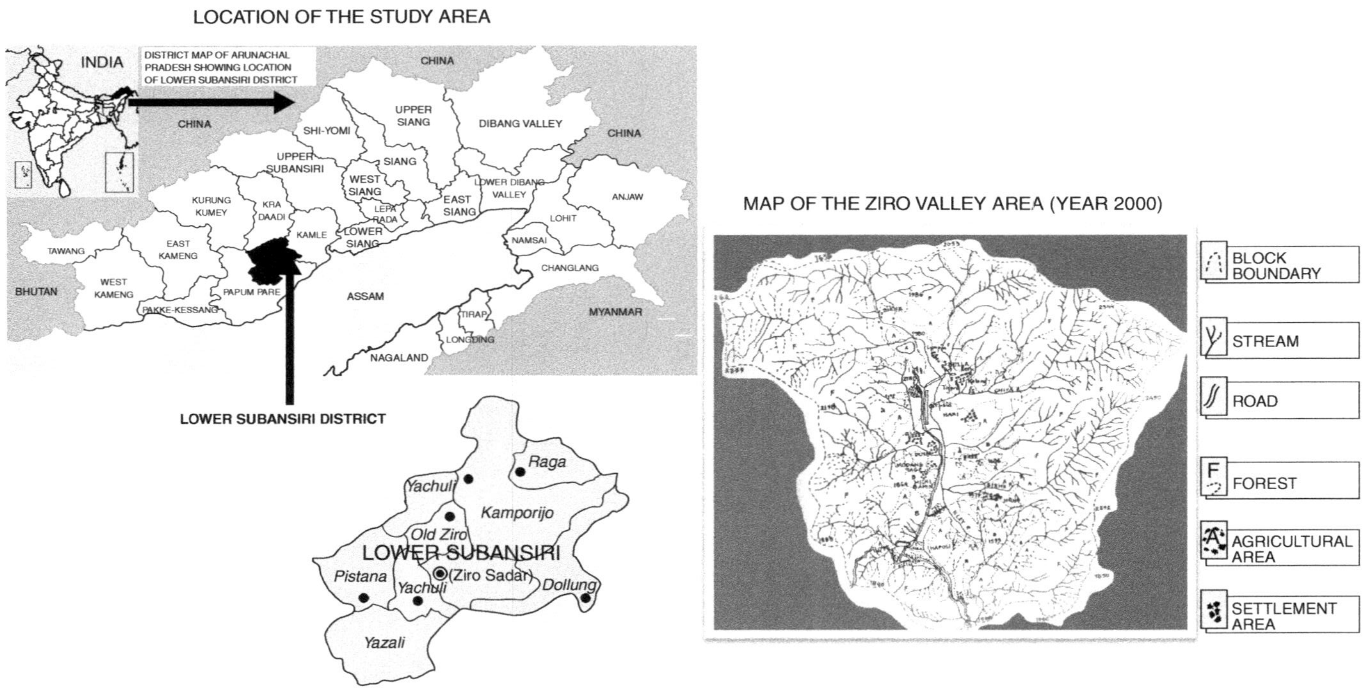

Figure 1.1 Map showing the location of the Ziro Valley and Lower Subansiri district in Arunachal Pradesh.
Source: Author.

a number of years after Independent India was formed, Arunachal Pradesh became a state of the Indian Union only in February 1987.

I had registered 'Khwab' in Itanagar to work towards a more sustainable way of life through bamboo craft development and its material propagation. This was at a time when the word 'sustainable' was not in fashion, at least not where I lived! Needless to say, in my loner gypsy-life mode and raw and determined twenties, I too was unfamiliar with this term in usage, but believe that I understood its importance.

So … once upon a time, in 1998, I decided to embark on a short journey that took me to the stunningly beautiful Apatani or Ziro Valley, located in the Lower Subansiri district of Arunachal Pradesh, which is actually a kind of plateau. At an altitude of 1564 m or 5000 ft above sea level, this valley plateau is the home of the unusually compelling Apatani civilisation. The elevation of the mountain surround that embraces the Valley lies between 1830 and 2900 m and the area enjoys a temperate climate with frost ridden freezing winters and heavy rainfall averaging 122 cm annually. A single mountain river, Kley, with low fertile alluvial banks, flows down through the Valley. The district headquarters, Ziro, is well connected by road to Itanagar and the neighbouring state of Assam and also has a kind of tiny airport then serviced by small aircraft and intermittent helicopter service and now basically non-functional.

I set off from my miniscule single garage room home in Itanagar, supported by my temporary stint with the G.B. Pant Institute there, which had agreed to fund my proposed research project based in Ziro, the unforgettably named headquarters of the Apatani area. Of course, for me it was a means to an end. My dream of setting up an upgraded bamboo craft-based livelihood and design development centre in Ziro remained unrealised, and this background research study that was to have enabled the dream to become a reality, remained just a dusty detailed research report in my library.

I connected to the Apatani people, a large and prominent tribal community of Arunachal Pradesh and partook of their history, amazing traditional knowledge, and picture-perfect landscape. One could never imagine looking at the natural perfection of the scenic Valley, that this was engineered by human hands (see Figure 1.2). A deeply shared love of bamboo endeared me to the community and cultivated the connection. I was to spend over three months in the Ziro Valley to complete my personal field research work related to the identification and utilisation of bamboo and cane resources and the traditional forest conservation practices of the Apatani. This intersected with the brilliant traditional land use system that had been followed by the tribe for centuries and has remained intact to this day. Almost every Apatani home has its own lovely bamboo garden interspersed with blue pine, oak and wild apple trees. To unravel their indigenous land use pattern and bamboo-dominated culture, my household surveys and related forest research and analysis was an education more useful and intensive than any university could have ever offered. The quantification of the actual utilisation of both the major natural resources was to have given a foundation for further work in the area. The amazing agricultural and forest conservation practices were to

Figure 1.2 View of the Ziro Valley from the Pyut More community forest, 1999.
Source: Author.

be documented as an inspirational learning for the rest of the country and the world at large. How important is experiential hands-on knowledge, how relevant and how enriching. How also neglected and packaged as traditional, outdated, and unscientific. How unfair.

The Tata Sumo that was to take me in supposed comfort from Itanagar to Ziro, the picturesque headquarters of the Lower Subansiri district of Arunachal Pradesh and the home of the Apatani community, was 45 minutes late. One of the unquestionably many curiosities in this land of '*khushi khushi*' (do as you like), the locally common humorous reference to our Arunachali ways, is that all local public modes of transport usually leave on the dot or not at all. The latter possibility was beginning to weigh heavily on my mind, when with a screech of tyres, a heavily laden Sumo turned the corner and skidded to a halt in front of me, an inch from my toes. Now in 2020, the Ziro plateau is a tourist destination with the nationally popular Ziro Music Festival, special tourist camp sites and pleasant village homestays. But back then in the nineties, the name and place were unheard of even in the neighbouring states of the east.

The Tata Sumo was the latest addition to the Arunachal public transport system and the fare of Rs 150 was reasonable by any standard even in 1998. Of course, one had to pay for it in the end as all such seemingly good things are mostly short-lived! There were ten passengers, including myself, and with all our respective luggage squeezed in, any illusions of a comfortable journey that I might have nurtured fled my mind instantly. Twenty kilometres before our destination, Ziro, it began to rain (as it usually does at some point in Arunachal) and the bulky belongings strapped onto the roof had to be accommodated inside. Thus, jammed tight in forcibly sandwiched camaraderie, we twisted and turned along the curving hill road in stoic silence. We reached Hapoli, the township that is now the new Ziro town, in a record time of 4 hours, which I would have greatly appreciated had I been less green in the face when I finally disembarked. The earth beneath my feet had never felt so good.

I settled into the near-freezing room at the Circuit House and was duly informed by an unusually sullen and unsympathetic caretaker that the heaters did not function and that coal was in short supply. After the way the morning had gone, I was almost expecting that. Dani Jenny, the local girl who was supposed to work as my field assistant, was nowhere to be found. This was not exactly my idea of an auspicious beginning and, so, with the winter chill seeping into my bones, I snuggled into the inadequate quilt and hoped for a better tomorrow. It had not been my day, and I wished for its rapid end. Besides, by then I was too exhausted to argue with Fate or anyone else.

Long rays of welcoming sunshine woke my chilled and aching body the following morning. Both Jenny and another local forest guide, Buru Loder from the nearby Mudang Tage village, arrived at the Circuit House soon after I finished my breakfast. Loder worked as a plant collector at the State Forest Research Institute at Itanagar, where I had met him and Dr Haridasan, an amazing ethno-botanist, who knew and loved Arunachal's flora better than anyone I have ever known. Loder had promised me assistance in my project as he was native to the Valley. Basking in the sun and looking out onto the township and beyond into the stunningly beautiful Valley with its rice terraces, pine plantations and bamboo gardens, we made our work plan for the week. My spirits lifted and a familiar excitement of meeting the new and unknown stirred my being. Things were looking up. Tomorrow was undoubtedly another day.

A Community Apart

A commonly known and oft-articulated historical fact indicates that the Apatani were continually hounded by the neighbouring martial Nyishi tribe and confined to the flat Valley, which they have inhabited for the past unknown number of ages. A peaceful community hemmed into the fertile valley plateau by their hostile neighbours, the Apatani evolved into a uniquely sustainable civilisation that learnt to survive expertly within constrained boundaries. Within their populous and tightly packed main ancestral village settlements, they developed a culture that is now globally recognised for its intrinsic conservationist disposition.

There are a total of 35 hamlets within the larger classification of the original seven villages, namely Hong, Hija, Mudang Tage, Michi Bamin,

Bulla, Dutta and Hari. All these villages are in a mindboggling radius of only 5 or 6 kms!! No other designated tribal community in north-eastern India is seen to inhabit such a limited area. In 1999, at the time of my field research study, the population of the Valley numbered only around 27,000. This, when pitted against the low 1961 census figure of 10,745 people, shows a constantly growing pressure on the Valley bowl, which is now currently about 50,000 strong (refer Government of Arunachal Pradesh portal 2021).

In linguist terms the Apatani language is referred to as a member of a sub-group of the Tibeto-Burman group of languages, where it is classified as belonging the Western Tani branch. Quite different from the Nyishi, Tagin, Galo, Adi and Mishing tribes with whom they share a common ancestry of sorts as part of this larger Tani group of tribes, the Apatani have a stand-alone ethos that is in complete ownership of its distinct cultural and land-use practices.

Apatani society is based on clans called *halu*, which is the foundational social unit of the tribe. Each of these clans forms a cluster, the traditional colony or *lemba* in the traditional settlement. Intense clan solidarity lies at the core of this close-knit community and enhances the social cohesion that is so crucial for its survival. Completely at variance with their neighbours, the Nyishi and Tagin tribes, the Apatani have a system of inter-village ritual exchange of gifts and labour that strongly bind and unify the entire village community. This amazingly remains unchanged since witnessed accounts of the 1940s by anthropologists like Christoph von Furer-Haimendorf, which are the first written record of the tribe. Nuclear families are the norm, and marriage is either by choice or sometimes arranged by parents. Marriages within a clan are prohibited.

Socio-political Traditions

The *Bulyañ*, or Village Council, is the powerful main governing body of the tribe that enforces and upholds law and order, manages community resources and settles disputes in the settlements. These are all administered according to tribal customary laws which continue to operate in the state as designated and protected by the Indian Constitution. This all-male body of clan representatives was traditionally made up of men of wealth and status but apparently did not serve as a governing head of the community in its ancient traditional avatar. Women did not find a place in the traditional Village Council. Traditionally, fines were paid in terms of *mithun* (Indian bison) and pigs and eighty-odd years ago the former were the main currency or medium of exchange through which land and valuables were purchased. Grain was also a barter item. This has obviously changed! There is now a headman at the helm of governance as appointed by the local state government.

Being a traditionally patrilineal society, Apatani women are completely excluded from the inheritance of all ancestral family land or immovable property. They only receive traditionally prized semi-precious stone jewellery from their mothers as part of their wedding trousseau. Of late there have been strong demands for equal land ownership rights and modification of customary laws by various women's groups and organisations. In general, however, women manage all household affairs and have an equal economic share of responsibilities within the household. The larger share of inheritance goes to the eldest son who succeeds his father as family head, while the remaining share of the property is distributed amongst his male siblings. This system of primogeniture still prevails but only applies to ancestral property and not self-purchased land holdings, which can be gifted at will or equally divided amongst brothers. If there is no male heir, the property passes to the closest male relative.

The Apatani have always had a head for economy and trade, which has led to the sprouting of new entrepreneurs amongst them. They also constitute a proportionately substantial number of government servants in the state. The emergence of a large breed of professionals has led to a slight lessening of agricultural activities and inevitably affected the traditional well-knit community structure. However, from all the interactions that I have had with members of the community, it seems that the recognition of change and need to preserve the unified strength of the community is still apparent within the tribe.

"Rice was Brought in the Dog's Ear and Cotton was Brought by a Bird"

The Apatani are an agrarian people and have followed a settled pattern of agriculture for as long as memory serves; in stark contrast to their immediate neighbours, the Nyishi community, who practice *jhuming*, or shifting agriculture. They have always depended for their housing and food needs on a well-developed agricultural system.

"It is said that rice was brought in the dog's ear and cotton in the bird's beak". Rice, bamboo, cane and cotton all have fascinating fabled origins. As always, seated around a heart-warming fire in Mudang Tage village, I was immersed in the mesmerising folklore of the Apatani; narrated with great gusto by yet another elder who was obviously a born storyteller. Each tale underlined the strong bonding and interconnectedness of all living beings. These endearing folktales of old, shared by community elders, have always been hearth stories throughout my early life. I have sat at many hearth fires (including my own), and I believe my strong association with the hearth relates to the learning I imbibed around its soothing warmth. Whenever I recall these hearth moments to memory, they envelop me in a scent of nostalgia and timeless flavour that embody the romance of a lost age. A home without a hearth has never really felt like a home for me and that rings true for many people who grew up around a hearth fire or, like me, got used to one in later years.

No cotton is grown, and there is no tradition of this as well, I am told. Raw cotton was always bought or bartered from the Nyishi tribe and then spun and woven in the Apatani home. It was never grown in the Valley as the staple food crop of rice remained the priority in the limited land area available for farming. *Tapo* is the wooden spinning instrument finished at one end with the seed of a tree, which was used in every Apatani house to spin yarn. In yesteryear, a host of traditional natural dyes from the forest were used to colour the yarn. This vegetable-dyed yarn has now been largely replaced by market yarn but is still woven by Apatani women into traditional apparel on the loin loom. Indigo, however, is grown for its blue dye, which is a part of the intricate design weave of the lovely traditional shawl and the highly attractive black and white sleeveless jacket (*jikhe tarw*). Leaves of the *Sankhw'* tree are boiled along with the yarn for 12 hours to get a lovely dark yellow colour dye. Dark blue dye comes from the mashed leaves of the *Mobu* tree, which are also boiled with the yarn for 12 hours at a stretch. After boiling, the yarn is left in the mixture for three days to get a proper colour. The process is sometimes repeated to get a better and stronger shade of colour. The *Mobu* tree grows only in the adjoining region, inhabited by the Nyishi tribe, so the leaves are sourced from there. *Pyai kencha* is the traditional natural black dye. *Pyai* is the name of a local tree, and *kencha* is 'black earth or mud', which is also found only in the areas that are home to the Nyishi tribe. Both the mashed leaves and mud are mixed and boiled with the yarn for the same 12-hour period and then left to cool to get the desired colour.

All things were natural (and, of course, what is now referred to as eco-friendly), in the Apatani home, until about twenty odd years ago. At the time, traditional utensils for cooking were made of clay, water was stored in dried gourd containers and drunk out of bamboo cups. Food was eaten and stored in flattened bamboo sheath containers. These wonderful plates were beautifully crafted, with their rounded ends finished and decoratively sewn with cane twine. All pottery was fashioned by hand, had an after-firing blackish colouring and was practised solely by the residents of Michi Bamin village in the Valley. However black mud or earth was never used for this craft.

In the late spring of 1998, I encountered another one-of-a-kind extremely fascinating article in the house of the then 72-year-old Duyu Tamo, the Head Gaonbura of Bulla village. It was a rusty staff referred to as the 'Hwrañ Yaju', with many free threads at its head, it is believed to have been in existence since the time of creation on earth of the first Apatani man (Abo Tani)! This nondescript simple looking staff is worshipped and said to cure the incurable and paradoxically have an equivalent evil 'avatar' if tampered with. A single thread from its crown is given to a worshipper or patient, as a cure or blessing. I was told that it flees from a burning house and later returns of its own accord to the new one built in its place! There is one other like it, a mystical comb, in Tajang village called Ranw Akhw.

Fire tongs or *mwge'*, one-inch-wide U-shaped fire-bent bamboo strips made of either planted or forest bamboo, found a place at every hearth. Bows and arrows were made from the forest bamboo, called *yayi*, but these too are now more or less a thing of the past. Hunting is now equally rare and an annual event, using more of bullet and gun than the erstwhile bow and arrow. Spears made of bamboo and wood are always kept at hand, lodged in a loft shelf near the entrance door, for dealing with an emergency or surprise attack. Although threats of this nature are also history, a hangover tradition continues to be followed in practically every household!!

'Donyi-Polo': An Enduring Faith

Danyi or *Donyi* (in Adi)[1] is the Sun in the Apatani language, and *Polo* or *Pwlo* is the Moon. This duo of supreme forces lies at the helm of the spiritual belief system and their fundamental role in human existence, and the creation of the earth's bounty is understood in un-idolised worship and ritual practice. *Donyi-Polo* or *Danyi-Pwlo*, is the indigenous nature-based animistic faith of the Apanti tribe, which hosts a vast pantheon of unseen spirits (*Úì*) that govern land, house and forest. They are considered to be both malevolent and benevolent as their fancy takes them, or so it is said! Their places of residence are believed to be in the natural environs, and they also cohabit human space. All manner of life-sustaining activities like farming, construction and hunting therefore require their approval and blessing for which numerous rituals and sacrifices abound. All of them require constant appeasement and share human-like desires and needs. They are envisaged as supernatural beings that have a human form and many *nyibu* or shaman- like priests of the faith, claim to have seen them in their dreams. These invisible spirits are believed to inhabit the natural environs and cohabit with humans. Various life sustaining activities like farming, construction and hunting therefore require their constant approval and blessing, for which rituals and sacrifices abound.

The duality of male and female counterparts is a common thread that runs through the many deities to be found in the faith. In usual stark contrast to their neighbours and, indeed, to most other Indian tribes of the subcontinent. The Apatani have a male Creator of the Earth, namely *Chañtuñ* with his counterpart – a female Creator of the Sky, namely *Didun*. The first human being or *Abo Tani* has an origin that is attributed to a deity called *Hilo*. These nature-bound traditions have further strengthened a harmonious coexistence with nature.

Twenty-five years and more have passed since the Donyi-Polo faith, followed by many tribal communities of the state (Abotani group of tribes), was included and recognised as a world religion. More recently, following in the footsteps of most of the indigenous tribal communities of India, there are a number of Apatani who have embraced Christianity; a faith which has steadily gained popularity, especially amongst the younger members of the community.

March to April heralds the warm onset of spring and is a time of celebration. It marks the beginning of the sowing of the rice and millet crop. At this time, starting at the beginning of March, Myoko, the main spring festival, almost a month long, is celebrated, finally ending in the first week of April. At this time every year, the entire tribe gathers together in a single village to celebrate Myoko with two whole days of feasting. In 1998, during my field study, I was a lucky witness to this festival, which was hosted by Hong village that particular year (See Figure 1.3). A different village is designated as host each year. This ceremonial inter-village connect nurtures the amity and peace that unifies the entire tribe. The festival cements age-old ceremonial hereditary friendships and involves rites for a better harvest with sacrificial ritual offerings to protect the newly sown crop from hail, pests, disease, and wild animals. An invocation of the spirits of earth and forest to enhance the fertility of the land and ensure the wellbeing of individuals or the community as a whole, forms a major part of the festivities. Once all have been fed and watered in the first two days, specially singled out friends receive extra favour by being plied with the choicest meat portions and specially home-brewed rice wine. These differently specified ceremonial relationships refer to particular familial friendships that have continued down through generations and are traditionally referred to as *bunwñ-ajiñ* and *manyañ*. The *bunwñ-ajiñ* is a friend from the tribe, while *manyañ* refers to a friend from the adjacent Nyishi tribe. These differently classified special relationships that are seen to strengthen social bonding have consciously unified the tribe and nurtured the visibly strong community spirit that binds it.

During Myoko, a wooden pole-like structure adorned with bamboo tassels called '*Babo*' is constructed in front of the celebrating houses to welcome guests. Amidst much merriment, young boys and girls swing around this pole on cane ropes as part of the festivities, adding to the joy of community celebration. While in conversation with Punyo Nikang, a traditional priest or *nyibu* (who was also the *Gaonbura* or headman of Hong village) and his wife Punyo Yakha, I was told that if a person married outside the tribe, then he or she could not perform the Myoko prayer ceremony.

In Punyo Nikang's house I was also asked to sample the famous Apatani '*tapyo*', made from the extract of a leafy forest fern, while I noted down all these incredibly fascinating details of the local culture. The *tapyo* is an acquired taste and made from the ashes of a fern which are soaked in water and then boiled. The residue is collected and dried. This infamous 'Apatani black salt' has a dull taste that is highly regarded and loved by the tribe but rarely finds favour with others! I regret to say that I was in the latter category! On the other hand, the pork and rice were delicious, and I was well able to make up for my earlier lack of enthusiasm by devouring large quantities of the same.

Other Apatani festivals, like Dree and Muruñ, are important traditions that further underline the strong spirituality and nature-based associations

Figure 1.3 Myoko Spring festival celebrations, Hong Village, 1999 (5 images as a collage).
Source: Author.

of the community, which tend to define the rituals and taboos held therein. In the rain-laden month of July, following closely on the heels of the Myoko festivities, the well-celebrated Dree festival involves prayers for the collective well-being of humanity and a bountiful harvest. On the other hand, the celebration of the Muruñ festival or tradition as it is sometimes referred to, is noticeably on the decline as it is largely dependent on rituals that are conducted only by the shaman-like priests (*nyibu*) of the Donyi Polo faith. The *nyibu* is recognised by his special shawl, called *jilañ*, which has a signature weave. His social designation is further endorsed by the particular cowrie shell adornment on his machete (in Apatani, and *dao* in *Assamese*) holder. The numbers of *nyibu* in the community are steadily on the decline, with the next generation being reluctant to follow this calling. A diminishing of the rites performed by them is therefore an obvious consequence. Stuart Blackburn, in his book *The Sun Rises*, which is a detailed study and translation of the ritual chants of Apatani socio-religious cultural practices, gives a very interesting comment on the Muruñ. He labels the Muruñ as a *twgo*, which he professes to be more of a ritual practice rather than a festival celebration as it is oft referred to now. The account below is an excerpt from his book, which is yet another strong indicator of the foundational ecological beliefs of the Tani (the Apatanis' preferred reference to themselves) people.

> *One old man who had already sponsored five Murungs said that he was planning another. When I asked him why, he answered without hesitation: 'Because without the spirits, there is no life. Without Danyi [sun] and the other spirits, we would have no rice, no growth. Without the earth, we can't live.' In these few words,this man, who was not a nyibu (a priest) and had little ritual knowledge, summed up the popular view of the Murung. A few men added that they hoped their Murung would remedy a specific problem. Some of them mentioned an illness, often mental or psychological, while others cited a childless wife, daughter or daughter-in-law, or infertility among animals. Still, even when the aim is to cure a specific problem, that condition is usually long-standing, and its remedy is considered part of a general wish to improve the family's fortunes.*
>
> (Stuart Blackburn. 2010. *The Sun Rises – A Shaman's Chant, Ritual Exchange and Fertility in the Apatani Valley*. Boston: Brill)

Note

1 Adi is a major tribe of East Siang district in Arunachal Pradesh and is also the language they speak.

2 Of Field and Forest

Figure 2.0 Image showing the traditional land-use pattern in the Ziro Valley with rice fields, tree plantations and community forest area, 1999.
Source: Author.

An Evergreen Legacy of Land Use

There is much to suggest that at one time in the remote past, this valley was a lake far above the gorges on either side and that the silt brought down by the streams from the surrounding mountains has filled out this lake and built up a plain, whose fertile soil has enabled the Apa Tanis to develop their peculiar type of agriculture and with it a settled form of life.

**(Christoph von Furer-Haimendorf, *The Apa Tanis and Their Neighbours*, Routledge and Kegan Paul. London: 1962)

DOI: 10.4324/9781003305538-2

It is hard to imagine a swamp once filled with amphibious reptiles (*buru*) when one looks at the breathtakingly beautiful Apatani landscape of field and forest embraced by green mountains rising to elevations of above 8,000 feet over the Valley bowl. Water and silt from streams running down these mountains have fed the once swampy marshland plateau and fertilised its well-used agricultural soil. These thickly forested mountains often have a snow-covered capping. The obviously tedious efforts of the Apatani ancestors who exterminated the *buru* and converted this swamp into its present-day format of well-watered rice fields, are laudable and quite mind-boggling. They channelled the mountain rivulets and built mini-dams to irrigate their rice terraces at will; all of which are carefully maintained and preserved as such today.

Land is, and always was, at a premium in the Valley and its scarcity has led to the currently extremely high monetary value of land in the region. Sale of land is therefore quite common in the Ziro Valley. This has remained unchanged for the past 75 to 100 odd years, in some sense, when *mithuns* (Indian bison also called *subu* in *Apatani*) and large grain baskets (*yagw*) were exchanged for farm plots. In the traditionally patrilineal Apatani society, a man's social status and wealth was designated by the size of his land holding from as far back as the forties, as confirmed by village elders who were present at the time.

Land has not been cadastrally surveyed, and there are no land revenue records in Arunachal Pradesh as a whole; with the exception of the three Jhumland regulations of 1947 (See Glossary), which recognise permanent, inheritance and transferable rights of the village communities and individuals over *jhum* lands (See Glossary), or lands under shifting cultivation

All community land dealings are dictated by customary tribal law, which continues to operate across the state and stands fully protected by the Indian constitution. Governed strongly by the *Bulyañ*, or traditional village council, which enforces these laws effectively and is responsible for the management of land and forest, the Apatani have retained their incredibly sustainable forest conservation practices and land use system which define their distinctive cultural identity.

Land is either individually owned or the joint property of the clan or is common village land. All bamboo gardens (*bije gonii*), cultivated land, homesteads with adjoining granaries and timber plantations (*sañsuñ*) are individually owned. On the other hand, the surrounding natural forest tracts (*more*), burial grounds and communal platform sites (*la'pañ*) are all clan owned, either singly or by two or more clans together. Village community land is minimal nowadays, and the Hapoli area has mostly been donated to the state government for its activities and use. This land was once confined to small grazing tracts at the forest foothill boundaries, which have now turned into sites for new houses or horticulture plantations. Since there is only the practice of permanent cultivation in the Ziro Valley, individuals retain property rights of inheritance and transfer through permanent occupation of their agrarian lands.

Agrarian Ingenuity

Confined to the 1058 sq km Valley, watered and fertilised by the Kley, a single small river flowing down from the surrounding mountains, the Apatani people have not wasted even one square inch of land. Of this total area, about a third, or 32 sq km, is cultivated land. Unlike other neighbouring tribes like the Tagin, Sulung (now referred to as Puroik) and Nyishi, which rely more on forest produce, the Apatani depend entirely on their own bamboo and timber plantations and permanent wet rice cultivation for all their daily food and housing requirements. Moreover, as they do not practice the usual shifting cultivation, or *jhuming* (see Glossary), their thick temperate forest cover has remained intact as compared to other areas of the state, where this is a customary practice. Extensive, well-watered rice plots (*aji*) cover a large part of the mostly level Valley and are watered by rain and a well-engineered irrigation channel system fed by the river Kley, which passes through their midst. A mosaic of wet rice fields bank either side of this river and are uniquely coupled with small circular fish ponds placed at intervals within their large expanse. The algae that grow in these well-placed fish ponds have nitrogen-fixing bacteria resulting in incredibly high organic or naturally grown rice yields per hectare. These harvest figures are supposedly comparable to, and even beyond, those figures of the not so "Green" any more "Revolution of Punjab State" in northern India, which is often referred to as the granary of India.

All the fields were traditionally ploughed by hand, and bullocks are not and were never used for ploughing. Livestock like the *mithun*, cattle, pigs and chicken are reared for meat and manure rather than for agricultural work. They are kept fenced in so as to protect the precious plantations and crops, which they fertilise with their excreta. Manuring or fertilisation of the soil is a great concern for the Apatani, who depend almost entirely on their own home-grown produce for their food. This has of course somewhat altered in the past ten years or so with new professions coming into the social demographic picture and agriculture becoming only one of many livelihoods. Composting of kitchen waste is a common practice, and the kitchen gardens are their main recipient. There has been a spurt in floriculture related to orchid farming and commercial fruit farming has emerged as a new income earner in the recent millennial years.

The gorgeous white blossoms of both the wild and the planted apple trees spring out of the exquisite Valley vista at intervals. Two indigenous pear (*pita*) varieties, one small and one large, add pink and white to the laden landscape. Orchard-grown plums and peaches (*takuñ*) are further additions to the local fruit basket. Many wild berries and fruit are eaten by the tribe and some, like the *salyo* pericarp whose flowers are eaten, also make a delicious flavourful chutney. The wild fruit and nuts gathered from the forest and enjoyed locally are described briefly here:

- **Pecha** is a wild apple tree that grows in the bamboo plantations and is not a commercially exploited species. Dried wild apple has a lovely sour flavour and has tremendous market potential in the natural and organic food markets.
- **Salyo** is the fruit peel of the *teeta champa* or Michelia tree, which is ground into a chutney. The flowers of this tree are also edible.
- **Semo** are red berries of the forest, oblong shaped and smaller than grape, with a slightly wine-like sweet-sour taste. They are most often found in the bamboo gardens and are also planted along the boundaries of the kitchen garden.
- **Bachiñ** is a yellow-green sour-tasting wild berry.
- **Sampvr** is a bittersweet wild blackberry variety.
- **Sañkhe** is a nut of the oak variety.
- **Kung** is a nut of the oak variety.
- **Hula'** is a grape like fruit with a sweet-sour flavour.
- **Ta'miñ** is a sweet and somewhat salty tasting wild fruit of a tree that has yellow-coloured wood.

It is, I believe, quite important to keep a record of edible wild fruit, especially berries that can be identified only locally and with local names. A changed lifestyle and external influences most often attack the food habits and local cuisine of smaller communities, and many tribal and mountain communities have lost their indigenous fruit and food to processed and marketed global foods.

Classifying and Cultivating Land the Apatani Way

Living in balanced and complete harmony with nature, the Apatani developed an intelligent classification of land use that fostered natural growth and maintained soil fertility over centuries (Figure 2.1). The scarcity of land prompted its environmentally sustainable and highly admirable use. The tightly packed settlements that make up the concentrated Apatani belt, all have a familiar topographical relationship with the surrounding forest (*more*), forest plantations (*sansuñ* or *salw*) and agricultural land (*aji*).

The high-level settlement area looks out over the low-lying rice terraces and is fringed by timber plantations of hardwood trees like oak and blue pine, called *sansuñ* or *salw* or *sadi*, kitchen plots and the famous bamboo gardens called *bije* gonii. These are strongly fenced in bamboo groves and most often interspersed with a few fruit and blue pine (*Pinus wallichina*) trees. Besides oak and blue pine, other hardwood species like *kwrasanw* (*Castonopsis*), *rwme* (*Alnus* species), *pwta'* (*Prunus* species), *semo* (*Prunus* species) are some of the Apatani plantation favourites, used for firewood, construction and craft purposes. Fruit trees like the peach (*takuñ)*, plum, wild sour green apple (*pecha*) and the *pwta*, a pear variety, are commonly grown in the bamboo gardens or near homestead land.

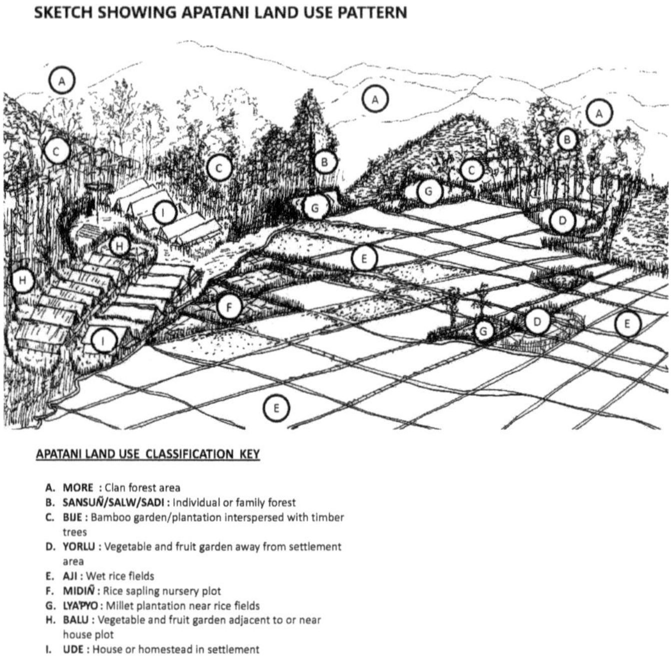

Figure 2.1 Sketch diagram showing traditional Apatani land-use classification in the Ziro Valley.

Source: Author.

Granaries are built individually for the storage of all the harvested crops or grain of the household. To protect these stilted constructions from fire, all the granaries are sited at the outskirts of the village, at some distance from the main huddle of almost wall-to-wall houses (Figure 2.2). The granary is used to store all the grain of the household and is a bamboo-and-timber based single room structure that is typically raised on stilts to protect the grain from dampness and being devoured by rodents.

In addition, there is an efficient distribution of land patches around the settlement constituting a variety of kitchen-garden plots. The one growing maize and vegetables is called the *yorlu*; a millet-only plot is called *lya'pyo* and yet another, namely *balu*, is the homestead kitchen garden located towards the rear of the house. Rice seedling nurseries, or *midiñ*, found edging the settlement and perched on higher level islands adjoining the rice fields, are yet another distinction in the well-conceived land-use pattern. Narrow pathways maze the settlement and connect all these to one another, finally culminating in the constantly flooded rice fields.

Figure 2.2 Traditional land-use pattern in Ziro Valley showing rice fields, millet plots, granaries, and bamboo and timber plantations (Collage of two images).

Source: Author.

The clear demarcations within the Apatani land-use system are listed and detailed below, except for the *bije* or bamboo gardens that have been case studied and are dealt with in detail in Chapter 3 ('Of Bamboo and Cane'):

- *more* / clan forest
- *sansuñ* /timber plantation
- *aji* / rice field
- *midiñ* /rice seedling nursery
- *bije*/bamboo garden
- *balu* /kitchen garden
- *yorlu* /vegetable maize plot
- *lya'pyo* / millet plot

More: Clan Forest

Forest tracts covering the surrounding hills, usually several hours walk away from the settlement, are owned and only used by a particular clan or jointly by two or three related clans that inhabit a particular village settlement. These tracts are not concentrated in a single block, but are dispersed throughout the Apatani area. In a particular forest area, only members belonging to the owner clan may hunt or collect cane, timber, firewood or any other forest product. In case anyone decides to violate this stringent rule, he or she will be liable to hefty and elaborate fines. In general, nowadays, hunting has declined considerably and is confined mostly to an annual jungle trip to snare the odd wild squirrel or bird for the Myoko festival.

The high-altitude rhododendron and orchid-laden mountain forest areas have a profusion of plant growth with a dominance of wild thorny bamboo species at many places. Tree species like Michelia or oak varieties, Guercus, Castonopsis, Prunus cornata, Betula, Schima wallichi, cinnamon and Taxus baccata are common in these forests. In the forests I visited, it was impossible to stretch out one's arms without encountering a thorny barricade! Forests are sacred for the Apatani. like the famous thousand-plus-year old Talle Valley sacred forest. Now designated as a wildlife sanctuary, it is said to be an extremely rare, naturally dying forest, as pointed out to me by a fellow traveller and American forester, and professor, Dr Michael Rechlin. Talle Valley has trees of immense girth and height, some of whose tops are invisible from the jungle interior. A comfortable 15km hike took me inside the Valley's fascinatingly wild interior, where the constant possibility of a leech bite was the only disturbing element in the otherwise breathtakingly beautiful and memorable forest experience. The sky was barely visible from inside the forest's depths and the girth of the oak was so unbelievably large as to be barely recognisable. Wild berries, fruit and nuts, along with the *Nwji* jungle leaves used in traditional animistic rituals, are collected at regular intervals by the tribe from the forest areas.

Sansuñ: Timber Plantations

The Apatani are conservationists of the best kind with a history of afforestation and well-maintained timber plantations as old as the tribe itself. In Hija village, I was proudly told by an elder (with heads concurringly nodding all around) that the Apatani traditionally planted a hundred trees for every tree they ever cut. This is a practice that dates back many thousands of years, he said, most definitely well before any government placards urged all right-thinking responsible citizens to do the same!

The statuesque blue pine (*pwsa*), which grows to heights of almost 170 feet, is not only a fuel and construction timber but seems to have tremendous symbolic and ceremonial significance. It is an integral part of the Myoko and Muruñ festivals, where it finds ritual use. Its resin is said to be a traditional herbal medicine for all manner of inflammatory ailments related to muscle, bone and joint injuries. Its long splits were also traditionally used as roofing cover but such roofs are now extinct and cannot be seen anywhere in the Valley nowadays. One or two such roofs were visible in the late nineties when I was around. Blue pine is often found within the bamboo garden and most often planted towards its peripheries, so that its felling or its shade does not disturb the growth of the planted culms. Its seedlings are planted pre-spring in the cold month of February at a distance of about 10 feet between saplings.

Aji: Rice Fields

The green-toned checkerboard of terraced naturally grown wet rice fields called *aji*, occupy the vast flat expanse of the Valley and extend towards the foothills of the forested mountain embrace. These fields are of two kinds: one that is kept permanently moist or under water, and the second that is left dry after the harvest. Of the two, the first is not ploughed and the crop stubble is left to rot and fertilise the soil. *Mipya*, *emo* and *pyapiñ* are the three main indigenous rice paddy varieties grown amongst the many others cultivated in the Valley. These are segregated and planted in different areas, having both early and late ripening varieties, and are the staple food of the community. Though men and women work in tandem in the rice fields, the bulk of the agricultural work was always, and still is, done by the women. Seedlings of the *emo* or indigenous late ripening rice variety are planted in between the stubble. In other cases, the field is cleaned and dug up before flooding it with water from irrigation channels. The soil is trodden upon to give it a thick, glutinous paste-like consistency and then planted with the early ripening rice varieties of *mipya* and *pyapiñ*. Fields that are close to the settlement area tend to be nutrient-rich, give higher yields and consequently attract a higher land price. These are all manured with a variety of organic waste in the form of pig and chicken droppings, cow dung and kitchen compost.

Midiñ: Rice Seedling Nursery

Rice seedlings are first planted in a nursery or *midiñ* in late February or early March. These nurseries are kept submerged under water throughout the year. Most of them lie close to the settlement on the fringes of the rice fields, to protect them from birds and to make it easier for the seedlings to be transplanted to the main fields. These plots are manured regularly with pig or chicken droppings and rich soil collections from beneath the stilted house and receives continuous vegetable and human bio-waste collected in pits below it. In the months before sowing, the water is drained out and the soil is kneaded by foot into a knee deep soft thick paste. Paddy seed from the granary is scattered over the surface and left uncovered. Water is allowed to filter in only after the appearance of green shoots. Transplanting of these shoots is done in April.

Balu: Kitchen Garden

Adjoining the homestead towards one side or at its rear end, is a small vegetable patch called *balu*. This patch is usually planted with vegetables or with a cane-like reed called *pepu*, which has nodes. It is woven into a spongy sleeping floor mat that is warmer than a bamboo mat and is always paired to form a soft mattress. This bamboo mat look-alike is laid all around the hearth for a cosy nap or seat. It is commonly used in every household and locally sold for Rs 200/-(2021) each.

Vegetables are grown mostly for home use, and only the surplus is sold in the local town markets in Old Ziro or Hapoli. The most commonly grown vegetables are chilli and a mustard variety (*giyañ hamañ*) that grows twelve months of the year. Other kitchen garden crops include pumpkin (*tape*), ginger (*taki*), soyabean (*po'tuñ peruñ*), French beans (*obyo' peruñ*), onion (*byaku'*), cucumber (*taku'*), marrow (*pinta*) and gourds. Potato, cabbage, cauliflower and tomato (*byayuñ*) are more recent additions to the Apatani palate, which traditionally favours meat and fish above all else and has limited vegetarian content. Chilli, mustard and coriander are the most common local market vegetables. Barring a few exceptions, ginger is rarely sold as it is considered sacred by the Apatani. These kitchen plots are as meticulously tended as the rice fields, fenced in with bamboo and fertilised with homemade manure.

Yorlu: Vegetable Maize Plot; and Lya'pyo: Millet Plot

The *yorlu* and *lya'pyo* are fenced-in plots, usually located in those areas that are unfit for wet rice cultivation. These plots are largely located on elevated mounds found interspersed within the flat expanse of the rice fields or on uneven sloping land plots found along the fringes of the rice bowl. The *yorlu* is basically a distant kitchen garden where crops like tobacco, millet (*sarse*) and maize (*tanyi'*) are planted along with fruit trees like peach, pear, plum or apple. Most of these trees are cultivated for home use.

The *lya'pyo* is primarily a millet-only patch. Millet (*sarse)* is solely used for brewing the local beer, which is mostly for home consumption and sometimes sold. Two kinds of millet are grown traditionally, an early ripening variety and a late ripening variety. Seedlings of both varieties are transplanted from the *balu* or *yorlu* gardens in the last half of April or early May. The early variety is harvested in August and the later one in early November after the rice (*emo* variety) harvest.

Due to the ever-increasing pressure of a growing population on the constrained land boundaries, fruit trees for home use are no longer planted as commonly as before, and some small agricultural plots have even been converted into residential sites for the construction of new houses. Moreover, there being no steady local market for the fruit, the surplus often goes to waste. Pear (*pwta'*) being the most common fruit tree in the Valley is always in surplus. In season, local pear prices fall so drastically that the fruit ends up being a banquet for the pigs! Other fruit trees like plum, peach (*takuñ*) and apple (*pecha*) can also be seen in home orchards. Transportation to a market in the hilly tracts is always a costly affair and not viable for most farmers logistically, even though road and air connectivity has improved hugely in the past ten odd years in the state. Today, many home orchards have been converted into either *bije* (bamboo) groves or kiwi and cardamom plantations, as some households have begun to run commercial orchards for income generation.

3 Of Bamboo and Cane

A Material Culture

The word 'Apatani', immediately conjures up the image of bamboo for those familiar with the tribe. "Bamboo is our life", an Apatani elder once said to me in 1999. This is completely true, even though its current use has diminished considerably today by comparison to that time. Such is the strong association with this natural material, that Apatani culture is often directly identified with the bamboo variety, *bije. Bije* or *Phyllostachys bambusoides*, is the smooth skinned, unusually erect, lovely bamboo that is traditionally planted and maintained by almost every Apatani household. These famed bamboo gardens (see Fig 3.0) are unique to the tribe, considered to be the only community in the world to have a signature male single species only bamboo-growing tradition of this kind and scale. They have for centuries nurtured their special brand of bamboo and blue pine, and a culture has evolved around these organic materials – a lifestyle that largely sustains them, even today.

> *"Amw Hintw Bunyi, bije mi Eli liha"*

An invocation chant of sorts, this line translates as, "*Amw Hintw Bunyi is* the female spirit that first planted the *bije* or bamboo".

An Apatani gentleman whom I interviewed in 1998 in Ziro, said that this dedication ritual called *Ude Payi* is done at the end of the construction of a traditional house, and that the chant is a blessing that ensures the durability and strength of the bamboo, timber and cane used in its construction. This *Ayu,* or chant, is sung like a song to appease the *Ude Úì* or any other malevolent spirits that may reside in the bamboo, timber and cane used to build the house, so that no evil, ill health or destruction befalls either the house or its residents. The *Ude Úì* is the House Spirit. All house-building defers to it and involves ritual appeasement with chicken sacrifices and dedications to ensure the good health and longevity of the house and its occupants. This invocation also forms a part of the Myoko and Muruñ festivals.

Bije (bamboo) was given by *Ayo Dopu Yaru* to *Abotani* or *Chañtuñ Tani*, the latter being the Apatani reference to the first human being or first recognised

DOI: 10.4324/9781003305538-3

Figure 3.0 View of a beautiful bamboo plantation of the monopodial Phyllostachys bambusoides (*bije*) species, 1999.

Source: Author.

ancestor of the tribe. According to the ancient Donyi-Polo nature-based faith, Dopu Yaru is a male spirit of the spirit world, who is said to have first brought bamboo to the Valley area, referred to locally as *myopiñ*. Its kinship with the tribe is practically singular as it was not traditionally seen, used or found elsewhere in the large Arunachal Pradesh state. It is this all-popular species of plantation bamboo (*Phyllostachys bambusoides*) that largely defines the cultural ethos of the Apatani, who use it in large measure for craft, construction, fencing and also as food (bamboo shoot). It is also the main component of the sacrificial ritual altars (see Figure 3.1) and further used to fashion some of the delicate bamboo shrine ornaments for all manner of traditional ritual practices during special occasions or festivals.

Similarly, cane too was believed to have been gifted to the Apatani ancestor by a female spirit of the same pantheon, called *Dimiñ Myori*, and then brought to the *myopiñ* (a reference to the Apatani forest area). All these fascinating accounts came from Danding Donyi of Mudang Tage village in the spring of 1999, as he sipped his tea intermittently and waited for me to get the correct spellings down in my precious (now yellowed but well preserved) spiral-bound notebook.

The material culture of the Apatani incorporates cane along with bamboo and it finds irreplaceable ritual use in festivals and some traditional ceremonies. Cane, as a natural resource, has almost the same importance as bamboo for the Apatani, who most often use both in conjunction with one another. The life of an average Apatani villager is unthinkable without these two materials. Both these versatile and wonderful natural resources are enshrined in a plethora of indigenous knowledge related to their growth and use.

Round bits of cane of almost 12 mm diameter and more also served as the unusual traditional nose ornament that adorned both nostrils of the Apatani woman (see Figure 3.2). The first piercing was done with a small sliver of bamboo that was later replaced by a cane disc, as the ear holes gradually enlarged. At this point there is an important need to correct a wrong popular notion that describes this ornamentation as a deliberate 'disfigurement' of Apatani women. This commonly held belief that has been all-pervasive, is ascribed to the fact that this so called 'disfigurement', was designed to protect the beautiful Apatani women from being taken off by raiding men from the Nyishi tribe in historical times. This is in fact diametrically opposite to the truth; this practice was in fact, an actual tradition of ornamentation much like tattooing, blue hair and the like and misinterpretation of its purpose basically comes from an outside and differing perception of beauty.

The use of cane as a preferred binding element, or rope, was primary in both vernacular Apatani craft and architecture and has been dealt with in more detail in Chapter 4. Cane's superior strength and durability in comparison with that of twine drawn from the bamboo culm, define its obviously

Figure 3.1 Ritual bamboo and cane ornaments and the bamboo and cane sacrificial altar.

Source: Author.

Figure 3.2 An elderly woman wearing traditional nose ornaments made of cane.
Source: Author.

discriminatory use in both joinery and construction of handcrafted structures. However, the difficulties in cane propagation and the daunting forest access of its natural sources, endow it with a higher cost and somewhat lower availability.

Of Bamboo: The Apatani '*Bije*' or Bamboo Gardens

The Apatani say that there can be no life for them without bamboo. Bamboo is a highly regarded, meticulously tended and jealously guarded resource. Of the nine types of bamboo found in the Apatani area, the Apatani largely depend upon their sole planted species of bamboo – *bije* or Phyllostachys bambusoides. Almost every Apatani home (about 98%) has its own bamboo plantation(s) of *bije*, which is wholly synonymous with the tribe. This is the only species planted by the tribe with the exception of the large diameter *Dendrocalamus hamiltonii* or *yayi*, the growing of which is so rare and the production so meagre as to be almost discounted as a plantation species. No other community on earth is found to parallel the kind

Figure 3.3 Collage of bamboo gardens Left a) Access pathway and boundary fencing between adjoining bamboo plantations Right b) A granary located in the bamboo plantation.

Source: Author.

of bamboo planting tradition or have comparable bamboo groves of the scale that the Apatani maintained and continue to maintain even now (see Figure 3.3).

Located on somewhat higher ground, at the fringes of the settlement area, a continuous row of well stocked bamboo gardens lined both sides of the main road leading into the village. These individually owned, carefully tended and strongly fenced groves are on an average about two or three acres, or one hectare, in area, the largest plot measuring about five acres. The gardens usually have granaries within their boundaries and are used primarily for construction of houses, craft work and fencing. They are typically interspersed with blue pine or oak (*kwra*) tree varieties, which basically provide construction timber or firewood. These are usually planted towards one corner of the plantation so that their felling has minimal impact on the precious bamboo.

There is a particular proportion in which the trees are planted in relation to the bamboo, which is said by the locals to be intuitive. However, on researching and subsequently quantifying the study numbers further; I found there was a fascinating and definite commonality of some kind of fixed proportion of the number of trees to the number of bamboo in the plantations. The number of blue pines must remain significantly low as compared to the bamboo so as not to hamper its growth. Wild apple (*pecha*) trees also tend to grow alongside the bamboo but are put to no commercial use as their fruit is dried and consumed locally. Traditionally, trees like the *sañkhe'* (Camellia cordata) and shrubs like *twmiñ/tamiñ* (Rubin cordifolia) whose leaves make wonderful natural dye, were planted here as well, but this practice has been discontinued with the introduction of commercial market-dyed yarn for weaving cloth. *Bije* is the lifeline of the community, which uses it in every sphere of daily existence from food to craft to housing in large numbers. All provided for by the slightly sloped bamboo garden that is most often interspersed with blue pine, other hardwood species and wild apple trees that symbiotically accentuate one another's growth and good health. Bamboo needs for annual housing and repair work are usually fully met by the household bamboo garden. However there were some houses that bought bamboo (*bije*) from within their own settlement or from a neighbouring village to fulfil their requirements entirely when building a new vernacular style house.

How to Plant, Tend and Harvest a Bije Garden Indigenous style

The *bije* plantations only have male bamboo which, since they do not flower or seed, cannot die! This bamboo species does not like standing water and must be planted on inclined ground to allow for sufficient drainage around its root system. Water-logging inversely affects the growth and size of the bamboo culm. Unusually this almost erect, high-altitude, monopodial species of bamboo is multifunctional and also has edible shoots. Its smooth long internodes encourages the high-quality basketry that the Apatani are famous for. Its high tensile strength, coupled with the fact that it is comparatively very resistant to borer attack, makes it the obvious favourite for construction. An orange powdery fungus that clings to its nodal area designates its age and consequent maturity of three years. This fungus is eaten raw or boiled by the tribe and is said to have medicinal properties related to digestive problems. After three years the culm is harvested for use in construction, an activity that necessarily requires mature bamboo.

The rhizhome shoots for a new bamboo garden are taken from an established garden and planted in early February. In a new garden the bamboo is planted from the top of the slope downwards. The root growth is said to be both better and quicker this way. The root of the planted shoot, or rhizhome, should be long to enable the growth of a larger diameter culm and quicker overall growth. The Apatani maintain that the longer the root length, the larger the diameter of the new culm. An entire one- or two-year-old bamboo culm is taken with the rhizhome and planted to allow for proper root growth. After three years, when culm maturity is fully attained, root growth stops. The planted culm is cut at a point, three or four internodes after leafing begins (Figure 3.4). A naturally occurring culm of this height is preferred for optimum growth. The culm is cut in the middle of the internode at an angle so that water does not collect and rot the bamboo. If it is cut too close to the node, the natural drying of the culm will spread to it and affect the node adversely. Traditional knowledge dictates that bamboo roots are laid out

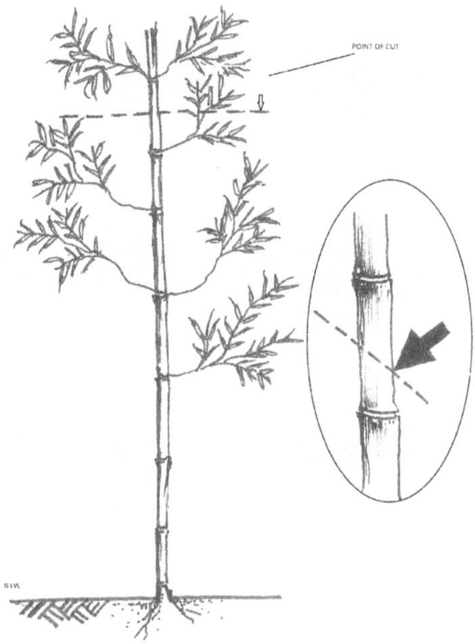

Figure 3.4 Diagram showing point of cut on the bamboo sapling for making a bamboo cutting for propagation.

Source: Author.

at preferred intervals of between 2.5 to 3 feet to ensure optimum coverage of the plantation area. Rigorous pruning of shoots is done to allow only one or two culms to grow from each shoot. Five years is the normal time it takes for a new bamboo grove to produce usable culms, and the entire coverage of an area of one hectare takes only three years.

Bamboo harvesting is only done between the dry months of September and February. This norm is strictly adhered to. Once planted, the bamboo requires little maintenance except that of constant vigilance to protect it from the mouth and feet of straying cattle or *mithun* (Indian bison). This is ensured by the elaborate bamboo fencing that surrounds each garden. Both these animals have a taste and fondness for the shoots and leaves of the bamboo and can destroy a plantation within minutes. Every year the gardens are fenced with the *bije* itself, using the entire length of the bamboo, including the tip of the culm and its leaf cover! Traditional knowledge also recommends the regular culling of 3-year-old matured bamboo from the plantation to ensure the proper and continual growth of new culms. An established garden will regenerate on its own and yield bamboo for a number of years.

Plantations under Study

In 1999, I sampled a few bamboo gardens as case studies to determine the density of standing bamboo and evaluate the productivity of an average plantation. Blue pine and bamboo plantations were studied and measured for the standing head count of both bamboo and pine. One typical bamboo garden was mapped and studied in detail while the density of five others was estimated by taking a number of sample circular plot areas, each measuring 19.6 sq m. The selection of the six case studies represented all the possible types of bamboo plantations found in the Ziro Valley. Hong Village had the largest area under bamboo cultivation.

Plot 1. Hong Village: Average-sized steeply sloping plot with both blue pine and oak trees.
Plot 2. Hong Village: Small-sized gently sloping plot with 2 blue pine trees.
Plot 3. Hong Village: Average-sized almost flat plot with wild apple, blue pine and hard wood tree species.
Plot 4. Hija Village: Small-sized gently sloping plot with fruit trees, blue pine, oak and other hardwood trees.
Plot 5. Hija Village: Very small-sized, gently sloping plot with only a few young blue pine trees.
Plot 6. Mudang Tage Village: Small-sized, gently sloping plot with some blue pine at one end of the plot.

Case Studies of Bamboo Plantations

OWNER: Takhe Gumbo
VILLAGE: Hong
PLANTATION AREA: 0.1 ha
DENSITY OF STANDING BAMBOO: 5100/ha
SAMPLE PLOT AREA: 117.6 sq m
NO. OF BAMBOO IN SAMPLE PLOT: 60

Takhe Gumbo owned four bamboo plantations, of which one of the largest was mapped (Table 3.1 and Fig 3.5) and studied in detail. Mapping was done by tracing a compass path along the boundary and taking readings at every turn in the path/circumambulation. Six sample circular plots of bamboo, each plot 19.6 sq m, were taken to assess the density of standing bamboo in the garden. A 100 per cent inventory was done on all trees over 4 cm in diameter.

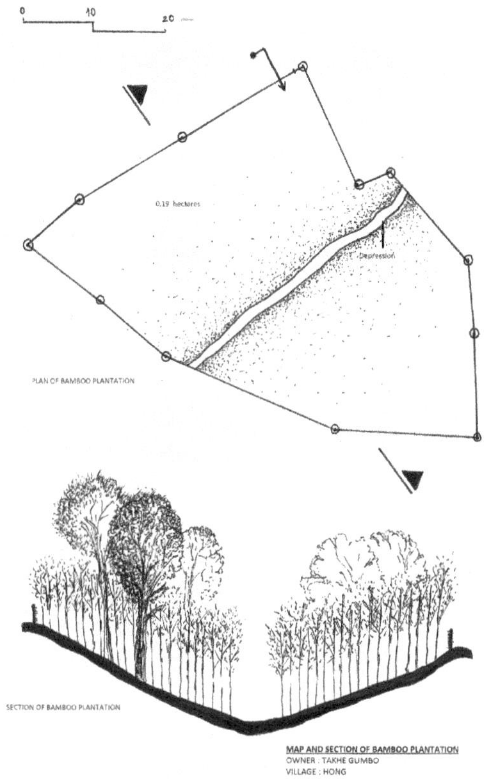

Figure 3.5 Sketch Map of Takhe Gumbo's bamboo garden, Hong Village, 1998. Source: Author.

The plantation had a number of scattered large-diameter blue pine trees with an understorey of hardwood trees and bamboo (*bije*). The hardwood trees were primarily oak varieties (*kra*) that were being harvested for house posts and to provide firewood. The upper storey was thin to allow for the proper growth of bamboo. Bamboo stocking was patchy as compared to the plots where bamboo was the primary planted species. Wherever large pine had been recently felled for housing purposes, there was a complete absence of bamboo. The total sample plot area of 117.6 sq m had 60 standing bamboo of which 34 were mature. The density of the standing bamboo stock varied greatly within the plot, the deep central valley between the two adjoining slopes having scarcely any bamboo. There was a total of 18 blue pine and 55 hardwood trees in the plantation.

OWNER: Takhe Gumbo
VILLAGE: Hong
PLANTATION AREA: 0.19 ha
TOTAL NUMBER OF BAMBOO CULMS: 969
DENSITY OF STANDING BAMBOO: 5100/ha
SAMPLE PLOT AREA: 117.6 sq m
NO. OF BAMBOO IN SAMPLE PLOT: 60

Takhe Gumbo's second garden was a small plot slightly less than 0.1 ha and had a slight gradient. It was a purely bamboo plantation with no trees. A single large blue pine stood on guard at one edge of the boundary. The bamboo stocking was admirable, with an average culm diameter of mature bamboo being 5 cm or 2 inches. The total sample plot area of 117.6 sq m had 60 standing bamboo of which 35 were mature. Some bamboo had recently been cut for housing repair.

OWNER: Kago Kamar
VILLAGE: Hong
PLANTATION AREA: Approximately 1 ha
DENSITY OF STANDING BAMBOO: 4900/ha
SAMPLE PLOT AREA: 98 sq m
NO. OF BAMBOO IN SAMPLE PLOT: 48

This sample plot area was in an almost-flat portion of a large garden, where there were no pine trees, and which had only a few wild apple trees that encircled the sample patch area. The total sample plot area of 98 sq m had 48 standing bamboo culms, of which 28 were mature, with diameters ranging from 4.5 cm to 6 cm. The bamboo stocking was uniform as compared to other studied plot areas and a few bamboo had been harvested recently from the plot.

OWNER: Dani Tabing
VILLAGE: Hija
PLANTATION AREA: 4228 sq m/1.04 acre
DENSITY OF STANDING BAMBOO: 4000/ha
SAMPLE PLOT AREA: 100 sq m
NO. OF BAMBOO IN SAMPLE PLOT: 40

This plot was not mapped. It was a combination of two adjacent bamboo plantations, one of which had been recently bought to increase the size of the existing plot. Quite a few bamboo culms, numbering 2,100, had been harvested in the recent month for rebuilding the owner's house, which had been completely gutted in an accidental fire a few months prior to the study (in December 1998). Due to this, the bamboo stock was low and interspersed with hardwood trees and pine. There were 13 large girth blue pine trees, 30 young blue pines, 17 hardwood trees of small girth, 15 large-girth oak trees and 8 scattered fruit trees, both pear and wild apple.

OWNER: Nending Rido
VILLAGE: Hija
PLANTATION AREA: Approximately 312 sq m/0.03 ha
DENSITY OF STANDING BAMBOO: 4500/ha
NO. OF STANDING BAMBOO: 140

This small plot area was almost rectangular and practically flat. A culm count of the total standing bamboo was therefore possible. The 312 sq m plantation had 140 standing bamboo and almost a thousand culms had been recently cut for house construction. The growth was uniform and undisturbed and most of the remaining bamboo in the plot area was immature. Productivity of the plot could not be estimated since most of the bamboo had been harvested. There were 6 very young blue pine trees of small diameter and there were also 4 granaries at one end of the plot.

OWNER: Buru Loder
VILLAGE: Modang Tage
PLANTATION AREA: Unknown
DENSITY OF STANDING BAMBOO: 4900/ha
SAMPLE PLOT AREA: 137.2 sq m
NO. OF BAMBOO IN SAMPLE PLOT: 67

This plot area had only bamboo and was located on an undulating mound adjacent to an expanse of rice field terraces. The total sample plot area of 137.2 sq m had 67 standing bamboo culms of which only 8 were mature. The bamboo stocking was uniform with a few bare patches near the path that led to the plot area. A large number of bamboo culms had been harvested in the previous month for both house and fencing repair work.

Important Conclusions and Observations (1999)

(*About an average established bamboo plantation from my case studies*)

A hectare can produce about 5,000 bamboo culms at a time. Of this number, 50 per cent were seen to be mature, implying that at any given time such a garden could yield about 2,500 to 3,000 mature bamboo annually. Since the average annual requirement was estimated at 393 culms per household, the above bamboo stocking was far more than what was required on an average at the time.

Many Kinds of Bamboo

In addition to the *bije*, there are a number of wild or forest bamboo that I was able to identify during my field research in 1998–1999, some of which have very specific traditional uses. The characteristics and types of local bamboo are summarised for quick reference in Table 3.1

A more detailed overview of each species is shown in Figure 3.5.

Bije

This is the king of all bamboo types, the only one planted by the Apatani and more indispensable for them today perhaps than in yesteryear. The community is entirely reliant on this monopodial bamboo, which lends itself to multiple uses, across housing, fencing, craft and as food. Its strength and durability coupled with its resistance to borer attack, makes it the obviously appropriate candidate for use in building construction. Its bamboo shoots must be eaten fresh, as they spoil quickly, but can be fermented or dried for longer use. It is almost unusually straight-bearing, smooth and even toned skin, high durability and short internodes makes it a high-priority species. It is easy to cut, not too heavy in carriage, and its tensile strength is recorded as being higher than that of mild steel. It is propagated vegetatively through the rhizhome by every Apatani household. In some cases, it was found to have replaced the kitchen vegetable garden, which had been converted into a bamboo plantation. *Bije* is oft referred to as a male bamboo, as there is no record of it ever having flowered. Up until the year 2000 a single culm sold for INR 3/- to 5/- onsite, which rose to INR 30/- in 2021.

Tabyo

Tabyo is the most commonly used wild forest species, but its use has gradually declined over the past twenty years. Its shoots are eaten by the *mithun*, so it can now only be found in the interior forest areas, from where carriage is cumbersome and dissuades its regular use. It is most often used for fencing rice fields (*aji*), millet gardens (*lya'pyo*), nurseries (*midiñ*) and vegetable and

Table 3.1 Types and characteristics of bamboo of the Apatani Valley

Species	*Local Name*	*Wild/Planted*	*Edible/ Non_Edible*	*AV. Outer Dia (in mm)*	*Av. Wall Thickness (in mm)*	*Internode Length*	*Local Usage*
Phyllostachys bambusoides	Bije	Planted	Edible	40–70	5–6	260–270	Construction, craft, twine/rope, fuel, bamboo shoot, ritualaltars, bamboo ornaments
Dendrocalamus hamiltonii	Yayi	Bought from the Nyishi tribe and sometimes planted	Edible (most commonly eaten both fresh and dry)	80–100	5–6	350–400	Craft (mugs),
Chimonobambusa callosa	Tabyo	Wild	Inedible	20–30	3–4	150–170	Rice field fencing, loft construction, craft, fuel
Chimonobambusa	Rwjañ	Wild	Inedible	10–25	5–6	140–150	Rarely used, fencing
Chimonobambusa	Tapyu	Wild	Edible	20–35	2–3	150–180	Not used any more. Was traditionally used to drive away birds and rats from farm produce.
Arundinaria	Tadvr	Wild	Inedible	5–9	2–3	400–500	Not used
Cephallostachium capitatum	Yabiñ	Wild	Edible, fermented bamboo shoot	25–35	2–3	480–530	Craft, loin loom component, bamboo shoot
Pleioplastus simonii	Hwbiñ	Wild	Inedible	35–45	2–3	400–500	Not used
**Unidentified	Tajvr	Wild	Inedible	8–15	Almost solid, sometimes with a tiny hole	800–900	For making shrine ornaments for the *mithun* sacrificial prayer

Figure 3.6 Collage of 3 different varieties of *Chimonobambusa* species of bamboo found in the Apatani Valley (Collage of three images).

Source: Author.

fruit gardens (*yorlu*), which are at some distance from settlements. Tabyo also finds uses in crafts, more particularly in basketry, and is used to make the bent component of the traditional hunting trap or *gwrw*. In Bulla Village *tabyo* was found to have been planted in some of the timber plantations and was used in the construction of the loft floor (*re'ke*).

It grows in clumps and has a distinctive thorny crown at its nodes. It is a dominant species in the interior forest areas, where I had some uncomfortably thorny encounters with it. Tabyo was seen growing in profusion with some cane and oak (*sañti)* species and effectively barring our passage through the forest. It is distinguished from other *Chimonobambusa* species (*rwjañ*) by its smooth outer surface and light reddish tinge. It has the largest diameter of the three *Chimonobambusa* varieties found across the Apatani Valley.

Rwjañ

This bamboo can be distinguished from the *tabyo* by the colour and texture of its outer skin, which is dark and rough in comparison. It is somewhat smaller in diameter and has a smaller inter-nodal length as compared to the two other thorny bamboo found locally. In the interior forest areas at higher altitudes, its growth and diameter were found to increase, and it tends to dominate forest vegetation. At lower altitudes, its growth is scattered, and its diameter small. It is rarely collected and typically abounds in deep forest areas where it grows adjacent to water sources like mountain rivulets or springs and shares space with oak trees. Its culms are monopodial and grow singly. Households that had small or no bamboo plantations were seen to use this bamboo for fencing their farm plots.

Tapyu

Another thorny bamboo that distinguishes itself by its distinctive perfume and slightly swollen thorny nodes, the *tapyu* also belongs to the Chimonobambusa family. The *tapyu* is quite like the *tabyo* but, unlike its typically flattened nodes, has contrastingly swollen and protruding nodal areas. Interestingly, its strong scent is not liked by either rats or birds, and so it is split into halves and placed in granaries and rice fields to keep them at bay! The younger, or immature, bamboo has an even stronger scent than its older counterpart, a scent that is said to derive from a white insect (unidentified species) that rooms inside the walls of the bamboo grove. *Tapyu*'s shoots are eaten locally, and the leaf and diameter size are akin to that of the *tabyo*. The *tapyu* is found at lower altitudes with high humidity conditions as in the Hapoli area and in the interior forest belt like the Hija village's *Pyut more*, or community forest area.

Yabiñ

Traditionally the *yabiñ* was the most important of all the forest species and found use in specific traditional craft items. Besides being an essential component of the loin loom, it was also used to make the *yapyo* or winnowing fan. Its long, lightweight internodes averaging 72 cm and smooth textural qualities made it extremely suitable for these purposes. Elders of Hija and Bulla villages told me that they recall its use in the construction of granaries, where it has now been replaced by the more durable and accessible *bije*. *Yabiñ*s shoots are locally reputed to make the tastiest and best fermented bamboo shoot (a fact that I can second by personal experience) and have a delicately sweetish flavour that is quite unique. This fermented version lasts for a year or more and is sold by the neighbouring Nyishi tribe per bamboo container (*sudu* in *Apatani* or *chunga* in Assamese) for INR 150/-(2021).

There are three locally known sub-types within this bamboo variety, which in itself is quite unusual: *yabiñ-bimbu, yabiñ-binkhe* and *yabiñ-to*. Each one of these, I was told, has different properties that encourage their varied uses. The first two varieties flowered and died in early 1998 and are said to have a traditionally known 45-year flowering cycle. The *yabiñ-bimbu* has the largest diameter size of the three, and the *yabiñ-to* has the longest internode (72 to 80 cm) and is used to make one of the components of the loin loom. The *yabiñ-to* is also poisonous when split and was traditionally used to snare wild prey in the traditional hunting trap. Fortunately, but strangely nevertheless, it chooses not to have the same effect on humans! Water-like liquid found inside this bamboo is used by the Apatani to cure stomach ailments such as dysentery and diarrhoea. So, a poison for one and a panacea for another! Its internode is 48 to 53 cm long. The sharp-edged split of the *yabiñ-bimbu* was traditionally used to cut the prenatal cord of a newborn baby. The flowering of this bamboo in 1998 led to its complete absence in the forest during the time I was doing my field research.

Tapiñ or Tajvr

This very often solid, very small diameter bamboo is naturally very hard and grows to extremely long lengths, typically bending and doubling over in imitation of a creeper. Its diameter varies from 8 mm (solid) to 15 mm with a very small hole, and its internodes are very long, often extending up to two-and-a-half feet and more.

This bamboo is considered sacred by the Apatani, who collect it once a year from the forest for both the Myoko and Muruñ festivals. It is only used for ritual practices and in the making of the outer weave cover of the ritual dried gourd container called *supuñ pinta*, which holds the rice beer. Small pieces of the *tapiñ* bamboo are finely split into a kind of flower form just above the node point and then tied with dyed wool to prevent it from splitting right through. Pieces of ginger (also considered sacred by the Apatani) are then pierced by the sharpened tips of this bamboo flower to form a ritual ornament. These bamboo stick ornaments are used to decorate the post where the sacrificial *mithun* or bison, is tied. This bamboo is sometimes (but rarely) bought from the Nyishi. *Tapiñ* or *tajvr* grows alongside the *tapyu*, at altitudes of between 5,500 and 6,000 feet at the foot of the forest fringe. If smoked, it is said to last forever and acquire an iron-like hardness, but otherwise it naturally rots in damp soil when exposed to it in the raw state.

Hwbiñ

The *hwbiñ* is a clump bamboo that is uncommon and can only be seen in the wild in parts of the Talle Valley area at altitudes of 2,400 m and thereabouts. I observed that it grows along with hemlock and fir. The variety of algae in its proximity indicated acidic soil conditions, which it seems to favour. It is not used locally and is very lightweight with long internodes, thin walls and diameters ranging from 2.7 to 3.7 cm.

Tadvr

This forest bamboo is found in the Talle Valley and Mani Polyang area, where it grows profusely in some parts of the forest. It seems to prefer altitudes of around 2,000 m, where it forms dense hedges 6 to 8 feet tall, growing in close partnership with oak tree species. Its diameter is a mere 1 cm or less, and it is very thin-walled, making it perfect for pen and straw making. It is not used at all by the Apatani.

Yayi

The *yayi* is a sympodial medium-sized bamboo that is not found in the wild in the Apatani Valley. It is only grown by some households in Bulla (Lempia) and Hija villages in their timber or bamboo plantations, and its bamboo shoots are commonly eaten. Otherwise, it is usually bought from the Nyishi traders in the form of bamboo shoot or as whole culms that are then used to make bows and arrows or beer mugs. Samples of these products can be seen in most houses, although their use is now very limited.

Delectable Bamboo Shoot

Bamboo shoot is a highly desirable and intrinsic part of traditional Apatani cuisine as it is for many other communities in India. Local cuisine flavours its meat, fish and green leaf dishes with well-made fermented (*hikhu'*) or dried bamboo shoot (*hi'*). Of course, the seasonal delicious fresh bamboo shoot vegetable is always in high demand as the Apatani do not have a large appetites for most other vegetables. All the villages were found to consume about the same estimated 14.5 kg per household quantity of bamboo shoot annually and 95 per cent of the households bought their requirement from the Nyishi tradesmen in the nineties.

Two varieties of wild bamboo are commonly used to make both dry and fermented bamboo shoot of which the slightly sweet flavourful *yabiñ* is much preferred, although it is less commonly available than the *yayi* (*Dendrocalamus hamiltonii*). The former is collected mostly from the clan or plantation forests of Hija and Bulla villages, while the latter, whose consumption rate is much higher, is entirely bought from the Nyishi neighbours at a cost of INR 100/- per cylindrical bamboo container (*sudu*), measuring 4 inches in diameter with a height of 20 inches, weighing approximately 450 to 500 gm. A dried bamboo shoot of the *yayi* or *yabiñ* costs INR 50/- for 250 gm or a small bamboo mug. Both fermented bamboo shoot *or hikhu'* of the *yayi* and *yabiñ* bamboo varieties costs INR 100/- for 500 gm (2021).

The average annual per household consumption of fermented bamboo shoot or *hikhu'* in the entire Valley was found to be about 4 kg, which could go up to a maximum of 18 kg in a single household, depending on the size of the family. In smaller households the consumption stood at an average of about 2.5 kg per household for the year. On the other hand, the average annual consumption per household of the dried bamboo shoot or *hi'* was much lower at an estimated 2.4 kg per household. In Hari Village an average of three basketfuls (*raju'*) of fresh bamboo shoot per household is collected during the season every year. Fresh bamboo shoot of the *bije* is collected three months of the year between March and May from the outside edges of the bamboo grove or its main path, while fermented bamboo shoot is available between the months of January and March. Quantification of this fresh *bije* bamboo shoot is quite impossible as it is just a surplus commodity, as and when found, since the actual bamboo stock was basically nurtured as culms for other far more important uses like fencing and construction.

Of Cane: In Search of the Elusive Cane

There are three varieties of cane – *tarpi* or *Plectocomia himalayana*, *tasvr* or *Calamus acanthospathus* and *takhe* or *twkhe* or *Calamus khasianus* – found and used on the Apatani plateau, of which two are endemic and one rare. All of these are primarily high-altitude cane varieties of varying diameter found in the wild (for details, refer to the next section), of which the first two are the most commonly used. When I planned an expedition to look for these cane types in the wild, I did so little realising that this search for the disappearing cane, was to become one of the most memorable journeys in my early adventurous and somewhat nomadic life.

Jenny belonged to Hija Village, which is one of the largest villages on the plateau, and she and I took a trip down there to talk to an elder, Nending Rido, who was a close clan relative of hers. It was decided in an animated meeting, that he would take us to his clan forest, called *Pyut more*, which had all three species of cane to be found in the Valley. *More* is the term used to describe the area under natural forest that belongs (by traditionally demarcated boundaries) to the entire clan community and cannot be encroached upon by any other clan member. A *more* is used by members of a particular clan for hunting, collecting cane and other forest produce. We decided to take off early in the day to begin the long and arduous trek into the forest from the village. The old man said it was about a four-hour continuously upward climb to the place where the elusive cane grew. Most species of cane are now quite difficult to find and even more difficult to regenerate, as they are habitat-specific and take about thirty years to mature. Interestingly, in recent times most Apatani men have almost abandoned hunting and rarely venture into the forest – the community having the largest number of men in government service in the entire state. Most of their cane requirement is bought from tradesmen of the neighbouring Nyishi tribe, who are more dependent on forest produce and are still expert hunters and use a great deal of forest produce.

The auto-rickshaw we had booked the previous evening, took us to Hija Village, 8 km away through the still dark and icy-cold morning. The elder looked not a little surprised to see us. He had been sceptical and apparently, doubting the seriousness of my intent and commitment regarding this unusual expedition. We repacked our sleeping bags and meagre rice and salt rations into two beautifully woven traditional cane backpacks (*le'ra*) provided by him and waited for Buru Loder, the fourth member of our expedition, to arrive. Our path led out of the village settlement and across the expansive rice fields, through the bamboo and pine plantations and up into the forest-draped mountain slope. The winter landscape stretched out in front of us, clothed in a gauzy white film of morning frost that added *wow!* to its stark dry beauty. In another three months, the white and brown would be transformed into a stunning lush green carpet with dashes of colour.

The pathway curved up into a narrow mountain track that completely disappeared into a thick brush and tree cover. The steep climb had begun. After a little while, the trees suddenly gave way to a chaotically wild profusion of thorny bamboo (*Chimonobambusa* species), the only bamboo species with a dangerously thorny nodal crown that pegged and tore at our clothes.

Species of large oak pushed through the maze of bamboo trying desperately to make their presence felt in its ever-spreading domination. At one point in our climb, we encountered patches of melting snow that indicated an altitude of over 8,000 feet, as told to us by our elderly guide. The Ziro Valley from where we had begun our journey now lay about 3,000 feet below us. These bald facts were designed to boost our, by now, slightly strained calves and flagging feet, and they cleverly did the job. Considerably cheered by having our efforts lauded and brought to light by our guide, we trudged upwards to a wholly unknown destination, eyes eagerly seeking the elusive cane.

Suddenly the steep path levelled out and we came upon a little bamboo bench with a conspicuous, single blue pine tree growing next to it. This was a *nyatu* in Apatani, or 'a place of rest' in the forest. Our guide, Mr Rido, explained that this resting place belonged to an ancient custom that, according to oral folk narrative, the Apatani had planted such a single blue pines at every rest point on their migration journey into the current Valley home. This route to the present settlement area and its point of origin is somewhat obscure. It is said to have been in three waves, but from a common origin. How long ago they migrated here is completely unknown, although there are clear anthropological indications* that their ancient home was a little further north, although not in the Tibetan highlands. What made all this so intriguing was that I discovered that neither the statuesque, often almost 150 feet tall blue pine (*Pinus wallichina*), nor the planted bamboo (*bije*) that is found interspersed with it, are native to the Valley area. *They are nowhere to be found in the immediate natural environment or even in the surrounding forests of the neighbouring region.* Mr Rido said that these '*nyatu*' were a regular feature to be found at various points in all the clan forest areas. However, my excited notion that perhaps one could embark on a search for *nyatu* points that would uncover the migration route to the present settlement site in the Valley was met with a dubious look. I still think it would make an amazing film story and dream of such an exploratory trek one day!

We stopped for a quick and welcome lunch of dried meat, bamboo shoot chutney and rice at the only stream we encountered on the entire long trek up the mountain. As we continued our journey going deeper and deeper into the forest recesses, the climb became steadily steeper and the path less obvious and increasingly obscured. The dense undergrowth prevented us from seeing beyond our noses throughout the expedition, and we plodded mindlessly in the footsteps of our guide. It had been almost nine hours, and my legs were beginning to protest strongly. Our forest guide, though well beyond the rest of us in age and experience, seemed the only amused one in the party and kept reassuring us that the end was near. Too true that, is what I felt. Suddenly there was a loud shout from him. Cane had been sighted. Our search had come to an end. It was like finding gold. Excitement seeped through my aching muscles, erasing the pain and taking over my numbed mind.

The obviously long unused 'campsite' (if you could call it that) was in shambles and needed an experienced eye to ascertain its unlikely origins! We therefore decided to make a 'new' one in a nearby clearing, amidst the tangle of cane and bamboo that filled the forest frame. As piping-hot black tea drained the tiredness from my limbs, I began to absorb the splendour of our surroundings. There were huge trees, the likes of which I had never in my life seen before. Their invisible tops were shuttered off by the dense profusion of uncomfortably thorny bamboo entwined with cane creepers that climbed up and into oblivion (see Figure 3.7). Outside the small camp clearing, we could not even stretch out half an arm's length without encountering a thorny obstacle. As we began to relax, Mr Rido cheerfully announced to all of us, that since I had wanted to spend a night in the jungle, he thought he would take us to the furthest point therein. Cane was to be found slightly lower down as well on the other side of the mountain. Catching sight of my expression at this revelation, he quickly and cleverly put in, with grand emphasis, that Jenny and I were the first women to have ever visited this spot or jungle. Traditionally, women did not come so far into forest depths, and they did no hunting. Further he added that most of the younger men of the village, too, had not made it this far. He had an amazing knack, I must say, for survival and getting back into favour. Ego-boosted and dead tired, we cooked our evening meal with gathered wild leaves supplementing the rice and dried meat rations that we had brought with us. There was water to be found in a spring close to the camp and with the indispensable *ilyo* or machete, Loder fashioned bamboo spoons and cups for us in no time. We all slept circled around a huge bonfire that burnt itself out under the open night sky. My last

Figure 3.7 Searching for cane: The author, with her companion, Ms Dani Jenny, and forest guide, Mr Nending Rido, during the Hija Village Pyut More clan forest trip, 1999.

Source: Buru Loder, Itanagar.

thought for the day was that it was far more comfortable than the icy and inhospitable Circuit House.

The next morning, I collected a few cane samples for identification and tried my utmost to take a few photographs through the shadowed dark green maze. I also discovered a huge jungle tick in between my toes, which had obviously fattened and feasted itself on my blood the entire night! Jungle souvenir. At least there were no leeches during this cold season. I had never developed a good relationship with them or found plausible reason for their co-existence on earth. I love most creatures of the earth, but you have to draw the line somewhere!

Making our way back down the impossibly steep incline, liberally sprinkled with thorny enemies at every turn, proved to be quite an ordeal. In a contradictory Murphy's Law sort of way, the downhill plays havoc with the knees and calves when steeply vertical. But mission accomplished, we could afford a leisurely pace. The last bit of foothill area and level path came as an anti-climax as our highlander knees continued in the unbroken bent mountain rhythm. Jenny and I parted company with the two men at the crossroads to the village and continued in robot mode on stiffening knees and thighs, towards the town. The last bus to Hapoli had long gone, and there was not a single auto rickshaw in sight in the scant dusky light. We walked on steadily towards our destination quite unmoved by this news. We were of the forest. We had conquered. Nothing so trivial could bother us anymore.

Varieties of Cane

The forest search for the three cane varieties (see Figure 3.8) was immensely satisfying personally, and the survey results of their massive consumption by the community completely mind boggling in scale when I recorded them in 1999!

Figure 3.8 Collage of different varieties of cane found in the Apatani Valley.
Source: Author.

Table 3.2 Types and characteristics of Cane found in the Apatani Valley

Species	*Local Name*	*Wild/ Planted*	*AV. Outer Dia (in mm)*	*Internodal Length*	*Local Usage*
Plectocomia himalayana	Tarpi	Wild	20		Most commonly used as twine in construction, craft and fencing
Calamus acanthospathus	Tasvr	Wild	10	150	Most important cane species for woven craftwork. Also used in traditional ritual practices.
Calamus khasianus	Twkhe/ Takhe	Wild	15–20	320	Rarely used. Used in craft and as twine in house interior work only.

All these cane varieties (Table 3.2 and Figure 3.8) are either collected from the forest or bought from the visiting Nyishi traders, who in turn source it from their own forest areas. Surplus cane is also sold locally by fellow villagers and very often a welcome gift of a coil of cane is offered to a fellow clansman. Many people have now started to plant cane saplings of *tarpi* or Plectocomia himalayana, along with the *bije* bamboo in their bamboo gardens with varied success. Hari and Hija villages were found to have the largest cane reserves in their respective clan forest areas, while Michi Bamin and Mudang Tage had no cane in their community forest areas at all. A few households in Michi Bamin village had also planted cane in their individual forest area (*sansuñ*). In most cases it was found that changing lifestyles that leave no time and space for cane collection and forest forays are the basic reason for high dependence on market cane and not really the actual availability of the resource.

Most cane types are difficult to propagate and often require a specific natural environment. Unlike bamboo, most durable and large-diameter species take almost thirty years to mature and, hence, are easily endangered. An average large-diameter cane propagates in only specific forest environs that are mostly inaccessible. Cane is a creeper unlike bamboo, which is botanically classified as a grass. It requires large trees to sustain its full-fledged growth. The smaller-diameter planted garden cane variety (see Figure 3.8) has limited usage and is of inferior quality but has an easier propagative ability. The peripheral forest areas are thus now completely devoid of cane due to over-exploitation. Jungle treks and forays into the forest interior, where cane is available, have become very uncommon nowadays due to the difficult terrain, access and changing lifestyles of the younger generation. This has led to over 60 per cent of the cane requirement being met from outside areas. Each Nyishi bundle (referred to in Assamese as *mutti)* contains 5 or 6 pieces of about 2 m of cane twine length of the *tarpi/tasvr* variety, as against the Apatani bundle that has between 12 to 15 pieces of 2 m length twine. The more-loaded Apatani bundle naturally therefore is more expensive than its Nyishi counterpart.

Tarpi (Plectocomia himalayana): *Tarpi* is the most commonly used cane in the Valley. Considered to be the most durable of all the three species, its twine widely used in housing, fencing and basketry. A slight slope and some shade are both essential for the proper propagation of this type of cane, which can grow well with pine, oak varieties and bamboo. In the wild it was seen to flourish with the *tabyo* bamboo (Chimonobambusa callosa) and a few large oak (*santi*) varieties. The leaf tip of this cane is typically elongated and the stem is thorny. It has the largest diameter, about 20 mm, among the three cane varieties found in the area.

Tasvr (Calamus acanthospathus): The *tasvr* is the second cane in line in terms of use and importance and is mostly a craft work cane, extensively used in traditional Apatani basketry. Three lengths of this whole cane is used to make the *subu-sa*, a tight thickly braided rope used to secure the sacrificial *mithun*. A 60 m average length of this rope is required once in three years by every Apatani household, since the Myoko festival is hosted by a particular group of villages every year. Most households make an annual pilgrimage to the forest to collect a coil of this cane during the Myoko and Muruñ festivals, where it finds use in the many ritual altars and ceremonies (Figure 3.7).

This cane species is found in the wild at altitudes ranging from 4,000 to 8,000 feet in subtropical forest zones. It has about 150 mm long internodes, an average diameter of 10 mm and a strength and flexibility that makes it highly suitable as a rim-strengthening element and for feet and side supports in basketry. *Tasvr* is used to fashion the traditional backpack or *le'ra* and the hard hat or *byopa* and in the fine cane binding of the gourd bottle or *supuñ-piñta*. Its sour, oblong shaped fruit with a white jelly-like filling has a vivid orange coloured textured shell cover and is always found at the top of the climber.

Takhe or ***twkhe***, ***(Calamus khasianus)***: The *takhe* or *twkhe* is rarely collected due its low durability in exposed conditions. It is sometimes used in the interior of houses as twine or in making the jungle backpack (*le'ra*). During the Myoko and Muruñ festivals it is sometimes used in ritual altars. It has an average diameter of 15 to 20 mm and a long inter-nodal length of 320 mm and is now endangered as a species and difficult to find even in the wild. Traditionally, it was used to make the famous large nose plugs earlier worn as ornaments by Apatani women.

4 Crafting Bamboo and Cane

Figure 4.0 Close linear cluster pattern of houses on both sides of a main village street in an Apatani settlement, 1999.

Source: Author.

DOI: 10.4324/9781003305538-4

The Apatani Settlement

No cadastral survey has been done in the Ziro Valley and, therefore, survey settlement maps are unavailable for study or reference. Yet, at a practical field research level, it is possible to identify and thereby assign a kind of typical pattern to the old settlement layouts. All the built-up areas of the settlements are concentrated in a huddle, with fenced boundaries between them, accentuating the funnelling effect of wind loads in the Apatani settlements (see Figure 4.1). The dense clustering of houses therein is usually linear, and all the houses face each other across a main central street or road that runs in between the two closely packed lines (see Figure 4.1).This 4 to 8 m wide main motorable road usually ends at a point in the vegetable gardens or at the rice fields after winding its way around the bamboo gardens found at the edges of the settlement. The other end typically leads towards the Ziro township or another neighbouring village. The tightness of the Apatani settlement structure on the Valley floor is indicative of a burgeoning population jammed into a confined boundary.

All settlements are physically divided into a number of wards or colonies called *lemba*, each defined by single clan occupancy. The *lemba* is the basic socio-physical unit that multiplies as clan clusters to form the community settlement or village. Every colony or *lemba* has a common open space where a *lapañ* or huge platform made of timber planking can be found.

Figure 4.1 Views of the traditional Apatani settlement showing cluster patterns, street and *lapañ* (community platform) layout and settlement structure (Collage of 4 images), 1999.

Source: Author.

These structures, where clan gatherings and meetings take place, are centrally located amidst a house cluster and can be found at intervals in the main street on land that is clan owned (Figure 4.1). When a *lapañ* is built, the site spirit has to be appeased through a sacrificial rite, and its blessing invoked for all those who use the platform. The *mithun*, or bison, sacrifice also takes place on the *lapañ*.

All the villages are electrified, connected to a piped water supply, and have basic amenities making them all self-sufficient units with well-defined boundaries. Typically, the built up area is concentrated at the core of the settlement, and the granaries and plantations surround it and lead outwards into the rice fields, which are again centrally located in the Valley area as a single acreage with multiple ownership.

A Bamboo-based Building Tradition: Vernacular Apatani Architecture

Until around the year 2000, 70 per cent of the Apatani houses used a very high proportion of bamboo in their construction, along with timber. Thereafter, there has been a noticeable decline in bamboo use in places where there is exposure to the elements, and bamboo's replacement with timber or brick is greater is now and very evident. Cane was used as a binding element in the form of twine in tandem with bamboo everywhere in traditional Apatani construction.

The Typical Vernacular Plan

It was difficult to distinguish between the vernacular houses as they all followed an identical plan and design that was basically a single large hall-like interior space. This was a typically narrow rectangular plan raised on timber stilts and an open verandah at the entrance (*byago*) and at the rear end (*uko*) (Figure 4.2). The 0.5 m to 1 m high raised or stilted structure allowed for the gradient of the land and was further braced with lateral timber or bamboo members.

The width of the average house ranged from about 3.5 m to 6 m, while its average length was in the vicinity of about 40 to 50 m but depended on the number of hearth fires within the structure. The *byago* and *uko* were open to sky, bamboo *machaang*-like platforms that extended outwards from the main covered built-up space of the house. They were used as sit outs, grain (rice and millet) drying and all manner of craft work like weaving, woodwork and bamboo cane basketry. One was greeted by an array of chicken cages suspended from a loft floor that partly overlooked a narrow semi-open space at the front of the house. This first room also housed the rice-pounding wooden ensemble. This further opened out into the long main hall of the house, which usually had two hearth fires, one at either end of its length. Bamboo mats (*pepu*) placed around the fire

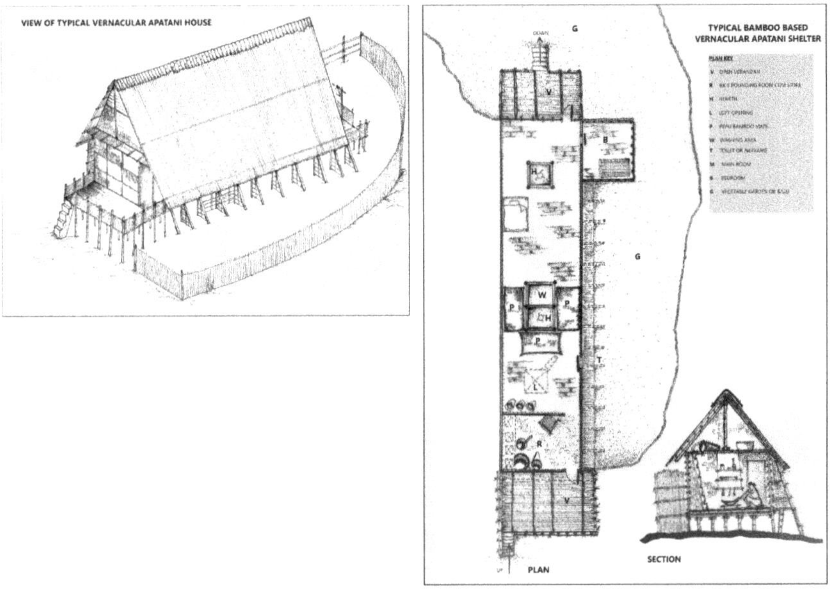

Figure 4.2 Plan, Section and 3D sketch view of a typical Apatani vernacular bamboo-based house (Collage of 2 images).

Source: Author.

served as sleeping and sitting mattresses. All basic household activities like cooking, sleeping and washing utensils were centred around these hearths. A three-tiered drying shelf of bamboo and cane was suspended over the fire whose smoke constantly dried the meat, recently harvested grains, and cane twine placed on or hung from it. A bamboo loft called *re'ke* extended across this hall and was used to store all kinds of bamboo and cane baskets, vessels, utensils, implements and other articles of daily use. A steeply sloping cut stepped bamboo pole or a single piece notched timber log ladder fitted into an opening in the loft ceiling and was the only access into the loft. Whole bamboo poles tied with cane twine formed the loft structure.

One narrow strip on a long side of the building was partitioned off into a toilet, which comprised a small opening in the floor that had a slop basin or bucket below from which the pigs housed under the structure cleaned out the contents acting as an efficient sewer system! The pigs were always enclosed within bamboo fencing underneath the house where their guttural voices regularly punctuated human conversation through the non-acoustic raised bamboo floor! The placement of the toilet in each house was such as to allow all the garbage or excreta thrown therein to drain into the side gutter-like alleys between the adjacent almost wall-to-wall houses.

Figure 4.3 The vernacular Apatani bamboo shelter (Collage of 4 images).

The vernacular house was poorly lit and had no windows. Its sole ventilation was provided by the two external doors at either end. Daylight hours being largely spent outdoors, this was not considered a necessity in the earlier context when all families were pure agriculturists. Various other daily or leisure activities like the drying of clothes, grain and vegetables, and craftwork and weaving were facilitated by the open verandahs in front and at the rear of the house. The house plot was typically fenced with split bamboo and normally included a small vegetable patch within its plot boundaries on one side or behind the house, if plot size allowed for it.

The Structural System

Bamboo, timber and cane from the surrounding environment were the only traditional structural construction materials used in building the highly eco-sensitive vernacular house. Due to the use of these lightweight materials, the traditional house was resistant to both earthquake and wind damage with very low risk to human life in catastrophe conditions. The timber stilts were deeply embedded into the ground and braced laterally with timber lengths or bamboo poles that strengthened and held the floor structure, which carried

Figure 4.4 Bamboo construction details of a typical vernacular Apatani bamboo-based house (Collage of 5 images).

Source: Author.

the bulk of the structural load in the traditional design. Very often these dually served as an enclosure for domesticated animals like dogs and pigs. Timber beams supported by these stilts crisscrossed to create the structural floor made of double-layered bamboo joists. A flooring (*myopu*) of flattened half split bamboo culm planks or timber planks laid alongside one another in lines and tightly secured with a single length of cane twine to the bamboo or timber floor framework that ran across the entire hall space. The verandahs and rice pounding room always had a different flooring, comprised of half-split bamboo lengths similarly tied down with cane twine. In some rare cases this flooring used the *tabyo* or wild bamboo, the only instance of which I saw in a house in Hong Village in the late nineties. The walls were constructed of bamboo mats secured to a bamboo or timber framework. The main vertical posts were usually of timber and supported the wall plates and beams while being fitted or lashed below to the timber floor structure. All the secondary supporting frame members were of bamboo. Thus, a large majority of the structural elements were of bamboo, including the rafters and purlins of the roof structure and the wall framework and cladding. End gable bracing was a common feature of the traditional Apatani roof. Doors were fashioned out of timber planks or made of bamboo matting framed in bamboo or timber (see Figure 4.4 for bamboo and cane construction details).

Gable roofs supported on a timber and bamboo framework were the norm, and the ridge pole was supported at regular intervals by vertical posts along its length and by transverse timber beams. In Hija Village, Yechi Yapii told

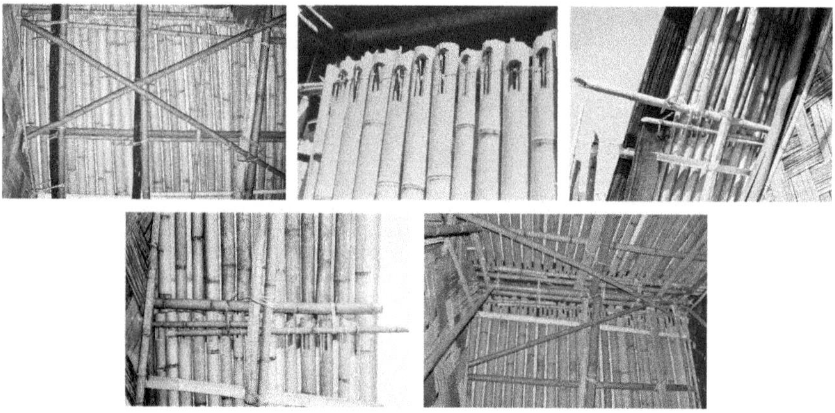

Figure 4.5 Collage of structural details of innovative bamboo tiled roofing in Apatani architecture (Collage of 5 images).

Source: Author.

me that the thatch roofs that were used earlier caught fire very easily and therefore had been replaced by ingeniously designed bamboo-tiled roofing panels. Although bamboo was equally prone to fire it was comparatively better than thatch and took longer to set ablaze! The untreated *bije* bamboo-tiled roofing had a life span of about 10 years and had become commonplace around 40 years ago (around the eighties I was told). About 50 per cent of the total bamboo used in the vernacular house (1063 bamboo culms) go into the making of this kind of bamboo-tiled roof. This was the bamboo quantification for the roof of an average 80 sq m area house.

The bamboo tiles were made into panels of 40 cm x 250 cm of half split bamboo lengths fitted into each other in long rows (refer Figure 4.4). Each bamboo tile panel was clamped between two small bamboo splits and joined together with a cane twine strand that passed through a hole drilled on top of each of the interlocking halved bamboo pieces (refer Figure 4.4). The roof structure below the tiles was always predominantly bamboo and was again cross braced with bamboo for resistance to wind and seismic pressure (refer Figure 4.4). Large-diameter bamboo was chosen to make these tile panels, and the ridging cover was made of flattened heat-bent bamboo culms (*mampo yamyo*) draped on either side of the ridge and secured tightly with cane twine. The bamboo planks used in the ridging of this kind of roofing needed more frequent replacement, as they were directly and constantly exposed to the elements. Before this kind of roofing became popular, thin pinewood bark plank roofing sheets were the norm, but are now not seen at all anywhere. The last sighting of pinewood roofing was in a few isolated cases in Hong Village 20 years ago, when I was living there. By 2000, Hari Village was laughingly called the *tina basti*, literally meaning 'tin village', as almost all the houses there adopted corrugated galvanised iron (CGI) sheet roofing after a devastating fire gutted the entire village in 1997.

Building Together as a Community

Traditionally, all clan members contributed voluntarily to the building or renovation of houses in the settlement as a norm. Building materials were collected from the bamboo and timber plantations and prepared over the dry winter, which enabled them to season naturally. In the traditional economy this system worked well, but in the changed context of the last fifty years, the vernacular house has lost both its economic and design viability. A lifestyle shift involving attitudes of privacy, school- going or college-going children, government interventions, new occupations and consequently newly perceived needs, all add up to a necessary change in construction design with longer durability and improved sanitation and ventilation quality. The tradition of community building has similarly become almost defunct in recent years, with the introduction of unfamiliar and imported building material; but the tradition continues in spirit with a common housing fund started by the community for the purposes of house construction. All members contribute to it and any of them can dip into it whenever required for an interest-free loan towards construction or renovation of their home.

Housing Issues

Low light and poor ventilation in the design of the vernacular Apatani house were perceived problems that can be easily remedied by the introduction of more openings within its design structure. The single hall plan is also no longer workable, as the new generation demands privacy and requires personal space! This needs minor design modifications, and partitioning of spaces within the existing structural design is possible. However, the small house plot areas with little room for ground expansion indicate a next-level expansion that has other major structural ramifications. Poor drainage in a high-rainfall area like that of the Valley, is consequent to water-logging and related unhygienic housing conditions in the congested Apatani settlements. This is further multiplied by the low gradient of the settlement sites. Waterborne diseases were a constant problem there in the nineties. The other housing issue was the extreme and omnipresent fire hazard with the organic materials-based house. This has been partially remedied by the local introduction of the much-aspired for fire-retardant CGI sheet roofing which, is in itself, a bad roofing option with poor sound and heat insulation qualities. This kind of roofing is not conducive to a constant and heavy rainfall zone with cold winter conditions. It denies both conversation and restful sleep to the inhabitants! Other options like asphalt shingle tiles and ceramic roofing tiles are yet to be introduced here.

Bamboo and Cane Utilisation in Housing and Fencing

Bamboo Utilisation in Vernacular Housing

There was literally bamboo in every inch of the vernacular Apatani home. The plantation bamboo *bije* was usually the sole species of bamboo used therein. There were some very rare cases of the use of *tabyo* or *Chimonobambusa callosa*, a locally found wild bamboo species that was used in tandem with *bije*. However, the use was minimal enough for it to be basically disregarded. The field survey in 1998–1999 indicated that the average new house consumed 2,500 to 3,000 mature bamboo culms while an average 1,500 to 2,000 similar bamboo poles were used in the renovation of this typical vernacular 80 sq m house. This kind of house had a life of between 8 to 12 years, after which it required rebuilding. The upgraded version of this house type was most often raised on cement concrete pedestals, had bamboo *tarza* or solid woven bamboo plank walls in a timber framework and corrugated galvanised iron roofing. In comparison to the earlier version, this kind of construction had a life of 30 to 45 years and its constituent elements did not need full or constant replacement. This introduction of new building materials, which are now used more frequently, has visibly led to fewer houses being built annually. The use of cement was, at the time, mostly confined to the raised pedestals or stilts upon which the house stood, or in the more rarely seen, plinth foundation. The only other element where cement found favour was in the construction of the steeply sloping steps that led to the front verandah (See Figure 4.3). Unlike today, where plastered walls and brick construction have made inroads into the architectural construct, cement was not seen elsewhere in the then, more commonly used, vernacular construct. The use of timber was also apparently extended to replace some of the external bamboo elements, which decayed far quicker when exposed to the elements. In 1999, the total annually average amount of bamboo used for housing (repair and new construction) in the rural Apatani belt was calculated to be a staggering 6,11,915 bamboo culms. At the time, an average of 192 houses were built every year. All of this bamboo came from individual plantations (Table 4.1).

Cane Utilisation in Vernacular Housing

The only species of cane used in traditional Apatani housing was *Plectocomia himalayana* or *tarpi* and it was calculated and sold locally in bundles of twine lengths. In a typical vernacular style house, the loft floor (*re'ke*) and the flooring (*myopu*) consumed the largest quantity of cane twine. Unlike bamboo, which could be reused in part, the cane quantity in a new construction and that in a renovated house was the same. This is due to the fact that all the cane twine is cut when dismantling the old house and cannot be reclaimed.

Table 4.1 Bamboo utilisation in a typical 80 sq m vernacular Apatani house (1999)

Village	*Ridging (Mampo Yamyo)*	*Roof (Pata')*	*Wall (Tarsi)*	*Floor (Myopu)*	*Loft Floor (Re'ke)*	*Verandahs (Byago & Uko)*	*Doors (Alye')*	*Pigpen (Parpw)*
Bulla	300	800	250	80	275	90	5	50
Dutta	300	800	300	90	300	100	7	60
Hari	300	773	250	75	250	100	5	150
Hija	300	690	300	90	300	100	7	100
Hong	300	820	250	85	300	100	7	150
Michi Bamin	300	800	300	80	400	85	5	60
Mudang Tage	300	700	250	70	250	90	5	75
TOTAL AVERAGE/HH	300	769	271	81	296	95	6	92

Table 4.2 Cane utilisation in a typical 80 sq m. new Apatani house (1999)

VILLAGE	*Number of HH's*	*Quantity of cane used in a typical 80 sq m House construction (in bundles of cane twine)*
Bulla	564	600
Dutta	40	500
Hari	268	500
Hija	372	600
Hong	483	500
Michi Bamin	107	500
Modang Tage	174	500
TOTAL	**2008**	**528 (Average qty)**

Note: One bundle (mutti in Assamese) of cane sold by a Nyishi trader has an average 12 m length of cane twine.

It was found that a total average of 528 bundles was used in the construction of a typically 80 sq m sized traditional house (Table 4.2). *An absolutely huge quantity of about 12,72,000 metres of tarpi cane twine, or an equivalent of 1,06,000 bundles of cane, was used annually for house building in 1999.* With an added annual figure of 9,13,450 metres of tarpi cane twine used in house repair, an estimated 21,85,450 metres of cane twine was used annually in the Valley as a whole at the time!

In a typical vernacular Apatani house (Table 4.3), the loft floor and the flooring consumed the largest quantity of cane twine. The amount of cane used in rebuilding a house or in an entirely new one remained the same, as all the cane twine was cut while dismantling the old house and thus could not be reused as was done with some of the old, reclaimed bamboo. About 7,200 m of cane length was used in the construction of an average 80 sq m house. *Tarpi* or was the only cane species used in housing and it was calculated and sold locally in terms of bundles or *muttis*.

Fencing

A staggering amount of bamboo culms was used for fencing, which was then redone entirely or repaired every year. An average 228 culms per year were used by a single household to effect this renovation. Every single plot of individually owned land, including the homestead area, had elaborate fencing of either *tabyo* or *bije* bamboo. Cane twine was the binding element in all manner of fencing, even in those that used a mix with timber poles. *Tarpi* was the only cane variety used for this purpose. Five lakh bundles amounting to a mind-boggling 51,14,750 m length of this cane twine along with an equally matching high number of 4,72,404 *bije* bamboo culms were used annually for fencing by all the settlements of the Valley (1999 field survey). Fencing done annually by a household consumed an average quantity of 274 bundles comprising 2,736 m of cane twine. About 300 bamboo culms covered the fencing

Table 4.3 Cane and bamboo utilisation in vernacular housing in the Apatani Valley (1999)

Village	*No. Of Households*	*Average No. Of Houses Built Every Year*	*Max. Average No. Of Houses Built Annually*	*Average Amount Of Bamboo Used per House (in culms)*	*Total Amount Of Bamboo Used Annually (in culms)*	*Average Qty Of Cane Used per House (in m)*	*Total Amount Of Cane Twine Used Annually (in m)*
Bulla	564	50	100	2000	1,00,000	7200	36,000
Dutta	40	3	8	1800	5400	6000	18,000
Hari	268	8	15	1500	12000	6000	48,000
Hija	372	50	100	2000	1,00,000	7200	3,60,000
Hong	483	60	80	2000	1,20,000	6000	3,60,000
Michi Bamin	107	15	30	2000	30,000	6000	90,000
MudangTage	174	6	10	2000	12,000	6000	36,000
TOTAL	2008	192	343	1900	3,91,400	6336	12,72,000

Figure 4.6 Bamboo fencing detail in a typical bamboo garden showing the use of bamboo poles with their leaf cover intact.

Source: Author.

of a one-hectare bamboo plantation plot as per my study at the time. This requirement has of course been reduced drastically nowadays as many people have started to replace traditional fencing with the less-aesthetic barbed wire and metal fencing that does not require regular maintenance or replacement, as did its predecessor.

The well-fenced-in bamboo gardens (*bije*) provided their own fencing material. This practically opaque and strong fencing design used the thinner upper leafy portions of the culm with its leaf cover, so that there was no waste of bamboo (refer Figure 4.6). This painstakingly crafted, elaborate bamboo fencing has been largely replaced by barbed wire in the past 10 years or so. Individual forest plantations or (*sansuñ*), and the separately located vegetable and fruit gardens (*yorlu*) are sometimes fenced with the wild *tabyo* (*Chimonobambusa callosa*) rather than the more commonly used and planted *bije* (*Phyllostachys bambusoides*).

Intricate Weaves: Traditional Bamboo and Cane Craft

> *'Craftwork or basketry was started by Abo Loma, the son of the first ancestor (Abotani) of the Apatani.'*

The above is yet another reference from the infinite Apatani folklore that came from another elderly acquaintance in Ziro, this time on the origins of

Figure 4.7 Male elders weaving a bamboo basket and a floor mat (Collage of 2 images). Source: Author.

basketry. Traditionally, bamboo and cane articles were made for home use by the male elders of the household (See Figure 4.7). Craftwork was never a means of livelihood or an economic activity in tribal society, for all products were made as essential articles for daily home use. As in every other ancient craft culture of the world, there was no formal system of passing basketry or other craft skills down the family, and the craft was learnt by constant observation and casual inexplicit guidance. In the current changed scenario with the formal non-contextual education system taking precedence over all else, these equations have changed, and skills are no longer being handed down to the younger generations. With each passing generation, these hand-skills are gradually and dangerously becoming extinct. Fine woven traditional basketry had its experts and required exceptional skills. The cane hat or *byopa* that rebuffed a bullet, the tightly woven *le'ra* that I backpacked with on my jungle treks, the intricately woven cane-bound gourd container or the *supuñ piñta* and the bamboo winnowing tray or *yapyo* that cleaned the rice grain are incredible, highly aesthetic, distinctive designs – ones that face extinction, as there is almost no one to make them nowadays. None of the practising master artisans are younger than 75 anymore and even 22 years ago, when I was living in Ziro, the use and availability of these specialized craft products was scarce.

A few male elders in some households still make and sell baskets and other bamboo articles locally within the village or in neighbouring villages. These items are made to order and are an income supplement at best. Some are given free of cost to the family and very often the cane used in their making is provided by the customer. The cane and bamboo species used in basketry and solid bamboo hand work are both strong and durable and specially selected for basketry and other craft work.

Attractive and intricate bamboo and cane basketry is the hallmark of Apatani craft, which is characterised by unusual shapes, beautiful design patterns and fine weaves. The traditional design needs little or no modification to market it. The perfection of form and function is replete in all of the huge array of traditional craft products. However, they are products of a different era, conforming to a lost context, where marketing naturally formed no part of the cultural ethos of the time or place. To adapt them to the now is only to change their use and size in some cases, which seems possible given that they all have an intrinsic flexibility when it comes to their multi-use. Additions of small design accents like zippers, magnet buttons, carved elements and similar minor design interventions may be another way of redefining these craft forms to ensure that they evolve and do not disappear.

Traditional Bamboo and Cane Craft Products

The Byopa *or Traditional Hat*

The helmet-like *byopa* worn by Apatani men is an absolute work of art. Exquisitely crafted in intricately woven cane, it is said to last a lifetime and is the forte of only a few remaining expert artisanal weavers. Conical in shape, it has a protruding beaklike hood extension rising upwards from one base end of the hat. The tight cane coiling that makes up the structure of the hat is begun from the crown and ends in strongly knotted stitch work. Made exclusively from *tasvr* cane, it is further adorned with hornbill feathers and beak. The hornbill beak protrudes outwards in front giving it its unusual identity. This headgear does not find many takers nowadays and is now worn only on festive occasions by some older men. Sadly, its makers are even rarer today.

The Le'ra *or Jungle Backpack*

The medium sized *le'ra* or traditional cane backpack was donned by the Apatani men on jungle trips and held basic food rations like the leaf-packed rice pouch, *tapyo*, salt, dried meat, and a few other small personal belongings. It consists of an entirely woven *tasvr* cane rectangular pouch which forks into two broad straps that further narrow into the shoulder straps. The straps have extensions of braided cane, one end of which is knotted to an attached cane loop at the base end of the backpack. A single strand of twisted cane twine hangs down the outer front face in readiness for the hunted game which is tied to it and strung along the back. It is an expertly woven and beautifully designed craft product. Having used it on multiple jungle trips, I am a diehard fan and can vouch for its unusual sturdiness, thankful lightweight and multi-usefulness that go well beyond its obvious aesthetic quality. The obvious ability that goes into its manufacture and the possibilities it presents in terms of modification are noteworthy.

The Yopo *or Grain Basket*

The *yopo* is a closed-weave basket made of either cane or bamboo and has a tightly braided cane string handle. It has the same square base, circular section design formula as the *yagw* but is more intricately woven. It shows close attention to detail and, as it is a ceremonial-use basket, great care is taken in its crafting. It is made in many assorted sizes as per requirement and originally was used to store rice powder or pounded rice grain. The cane version of this basket is called *yaso yopo* while the bamboo one is called *bije yopo* so named after the bamboo used in its construction.

The Chiba *or Rice Pouch*

A small pouch woven in cane, the *chiba* is used to hold cooked rice for a jungle trip or to carry a field snack during farming. It is essentially a tiffin carrier in pouch form with a woven flap and has a life of up to five years with regular use.

The Supuñ Piñta *or Gourd Bottle*

A gourd is often left to dry on the creeper itself and then cut open from one end when it hardens to form a naturally shaped bottle flask. The *piñta* or gourd used for this purpose is carefully selected, as its shape is crucial in the making and finishing of the final product. The *supuñ piñta* is a ceremonial flask used to keep and distribute rice beer during various ritual ceremonies, marriages and festivals. It consists of a dried gourd tightly covered with a layer of intricate woven cane that moulds it into the shape of a flask with base, lid and strap handle. One or more pieces of this highly prized item is found in every Apatani home. This amazing basketry technique lends itself to immense possibilities for design development and packaging.

The Rapya/Raju, *or Firewood Basket*

The firewood carrier basket called *rapya* or *raju* is an elongated, crudely woven basket with an open hexagonal weave. It is the most commonly used basket in every household and made entirely from *tarpi* cane. A double weave bamboo version of this basket, called *barju,* also exists but is not as typical or frequently used as the *raju*, which is far stronger for the job it needs to do.

The **Lwha'** *or Dried Food Basket*

The *lwha'* is a small basket with a square base and a circular section. Its four base legs and rim-strengthening elements are fashioned out of half splits of *tasvr* cane, while *tarpi* cane is preferred for its woven body. It is used to store the delicious dried meat and fish and lasts up to five years of regular daily use.

The **Paro' Patvr** *or Chicken Cage*

The chicken are housed in a pyramidal-shaped open-weave bamboo basket called the *paro' patvr*, which serves as both a coop and carrier for domestic poultry. A very typical craft item, it is usually made by a male member of the household for everyday use. This handcrafted slightly rough-woven piece was commonly found perched upon a bamboo loft at the entrance of the traditional Apatani house in the rice pounding room. It has a small opening and a small single-piece wooden door through which the chicken can enter or exit.

The **Pwpiñ** *or Floor Mat*

The *pwpiñ* is an almost square, tightly woven bamboo mat made from the *bije* bamboo. Paddy or millet is spread on it and sun dried. Its fine herring bone diagonal weave pattern often acquires a natural, multi-toned hue, which is highly attractive and formed from the different layers of bamboo skin used in its construction.

The **Sarse Pakhe** *or Bamboo Tray*

The *sarse pakhe* is a huge, shallow bamboo tray for drying the millet that is used to make millet beer. It is roughly woven out of bamboo and smoked over the hearth fire, which lends it a protective textural sheen. This is a very common product and a must in every Apatani home even nowadays.

The **Yatw'** *or Bamboo Raincoat*

Until about 25 years ago, when ugly plastic usurped its position, the traditional two-piece bamboo raincoat was common raingear during farming. This bamboo raincoat or *yatw* is usually carried on the back underneath the head basket or held by the forehead with cane rope. Its two components are held together with long twisted bamboo twine. The upper part is rectangular, with one end extending into a triangular cap that protects the head. The lower portion is also a rectangular piece in the same weave pattern and shields the back from heavy downpours. Both parts have braided cane straps that hold the contraption across the forehead. This ingenious design is in folding, and the upper part folds over the lower piece for ease of storage and carriage. The edges are strengthened and tightly bound together by half-split pieces of whole *tasvr* cane lengths. The entire piece is double-layered with each open hexagonal weave pattern in bamboo sandwiching a waterproofing of tree bark or dried leaves in between the two layers.

The **Yadiñ** *or Jewellery Basket*

The *yadiñ* has also lost its importance and use in the present context and is now very rarely made or even seen. It is a lovely piece of bamboo basketry with a tall cylindrical shape and slightly narrowed neck covered with a similarly woven lid. Personal possessions like jewellery and clothing were stored in this basket. Its rim elements and edging were fashioned from *tasvr* cane.

The **Yagw** *or Paddy Basket*

The paddy storage basket or *yagw* comes in varying sizes and is perhaps the most commonly found craft product in every home and the most important one. It is a closed-weave basket with a square base and circular rim. It is used for collecting and carrying the paddy from the fields. Wide woven straps interlaced in the side binders allow the basket to be supported on the forehead and carried on the back. The woven body is either of bamboo or cane, but the rim and side supporting pieces are as usual always made of the sturdy half-split *tasvr* cane. The weave at the rim is sandwiched between two coils of half-split cane, which are further bound together by cane binding that passes over and through the rim elements.

The **Yakhañ** *or Lidded Basket*

The *yakhañ* is a huge, four-foot tall lidded basket that is made of either cane or bamboo. It is a closed-weave cylindrical container used for storing millet beer (*o*). To allow for this wet storage, the insides are coated with a natural tree resin called *sañkhañ,* which derives from the *sañkhañ sanw* tree, found only in the Talle Valley area owned by Hong Village, and in the Nyishi belt. Smaller versions of this basket are sometimes made and can be seen in some villages. Its rim elements and side vertical support braces are as always fashioned from split *tasvr* cane.

The **Yapyo** *or Winnowing Fan*

The winnowing fan or *yapyo* is still a necessity and retains its popular use in every household. It requires expert craftsmanship and special skills due to its shape and design, and there are now only two or three people in every settlement able to make these. It is beautifully finished, strong, tightly woven and has a typically flared U form and a standard size. All four edges of the fan are strengthened and finished with split *tasvr* cane lengths. It is the only craft product that is exclusively woven from the wild *yabiñ* forest bamboo.

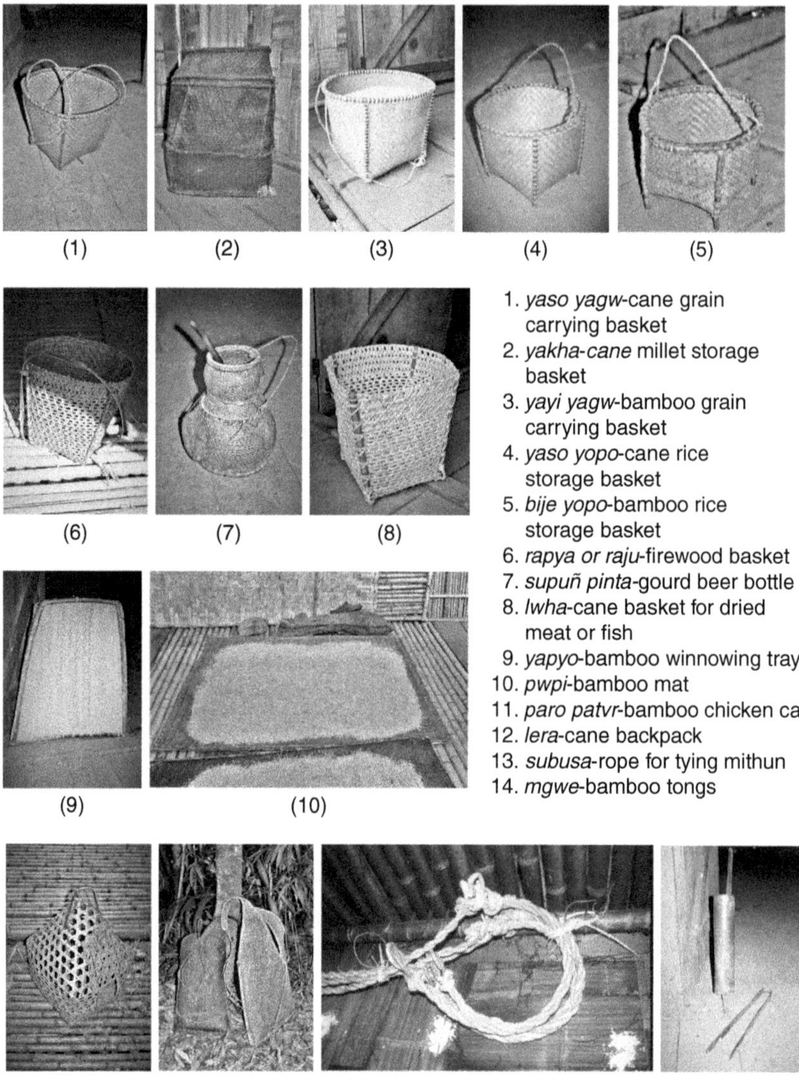

Figure 4.8 Collage of traditional bamboo and cane craft products with key giving descriptive and traditional names (Collage of 14 images).

Table 4.4 Bamboo and cane utilisation in traditional craft products

Product Specification	*Local Name*	*Size*	*Bamboo Type*	*Cane Type*	*Qty Of Bamboo Used (In Culms)*	*Qty Of Cane Used*
Cane grain carrying basket	Yaso yagw	Large Medium	-	PH+ CA	-	300 m P twine, 4 m whole CA 240 m twine, 3m whole CA
Bamboo grain carrying basket	Yayi yagw	Large Medium 44 cm dia x 35 cm ht	-	-	8 3 TO 4	3 m whole for rim
Cane grain storage basket	Yaso yopo	20 cm dia x 24 cm/ 75 cm ht	PB	CA	-	2.5 m whole
Bamboo grain storage basket	Bije Yopo	28 cm dia x 24 cm ht	PB	CA	2	3 m whole
Cane firewood carrying basket	Rapya/Raju	Standard	-	CA	-	117 m twine
Bamboo firewood carrying basket	Barju'	Standard	PB	CA	2	-
Cane meat/chilli/yeast container	Lwha'	Standard	-	CA +PH	-	33 m ph twine, 2 m whole CA
Bamboo lidded container for rice/ vegetables	Aji pwha/ Apiñ pwha	Standard	PB	CA	1	2 m whole CA
Cane lidded container for rice/ vegetables	Aji pwha/ Apiñ pwha	Standard	-	CA +PH	-	4 m whole Ph, 2m whole CA
Bamboo container for dried meat	Yo patvr	Standard	PB	-	5	-
Bamboo tray for drying millet	Sarse pakhe	Standard	PB	-	5	-
Rice container	Yadiñ	Standard	PB	-	2	-
Clothes box	Yadiñ	Standard	PB	-	7	-
Winnowing tray	Yapyo	Standard	CP	-	10	-
Gourd bottle	Supuñ piñta	Standard	PB	CA/PH	-	44 m twine
Huge storage basket for millet	Yakhañ	Large 32 cm dia x 50 cm ht	- PB	PH+CA	10	612 m twine

(*continued*)

Table 4.4 Cont.

Product Specification	*Local Name*	*Size*	*Bamboo Type*	*Cane Type*	*Qty Of Bamboo Used (In Culms)*	*Qty Of Cane Used*
Bamboo filter for millet beer	O sadvr	Standard	PB	PH+CA	1	
Cane filter for millet beer	O badvr	Standard	-	PH+CA	-	44 m twine
Bamboo mat	Pwpiñ	130 cm x 130 cm	PB		5 IF DOUBLE LAYERED	
Cane backpack	Le'ra	Standard	-	CA	-	20 m whole
Dao cover	Ilyo' Hubyu	Standard	PB	CA	-	5.5 m whole
Cane hat	Byopa	Standard	-	CA	-	10 m whole
Bamboo raincoat	Yatw'	Standard	PB	CA	4	3 m whole
Fishing trap	NgwiTakhuñ	Standard	PB	-	1= 8 PCS	-
Hunting trap	Gwrw	Standard	PB,CC	CA(negligible)	-	-
Tongs	Mwge'	Standard	PB	-	1= 200 PCS	-
Rope for tying mithun/ bison	Subu sa	1500 to 2000 cm	PB	CA	-	60 m whole
Basket straps	Se	Standard	PB	CA/PH	-	48 m twine
Hen coop	Paro' patvr	Standard	PB	-	1TO 2	-
Bamboo and cane mug	Turla'	75 cm dia x15 cm ht	DH	CA	1=6 PCS	6 m twine for binding only
Bow and arrows	Apw Alyi	Standard	CC, DH	-	1=150 PCS	-
Bamboo cooking spoon	Punyu	20 cm long	PB	-	1=100PCS	-
Chilli/salt container	Yormw/pila sudu	20 to 25 cm ht	PB	-	1=7PCS	-

PB: Phyllostachys bambusoides
DH: Dendrocalamus hamiltonii
CC: Chimonobambusa callusa
CA: Chimonobambusa acanthospathus
PH: Plectocomia himalayana

The Yopatvr

The *yopatvr* is now an obsolete craft item, but some remaining old samples are visible and gathering dust in a few isolated houses of Hong Village. This large suitcase-like shallow woven bamboo box was used to store dried meat, fish and fermented soyabean. It had a woven lid and looped clasp and was said to last for at least ten to fifteen years, depending on its regular usage. Unfortunately, it is no longer made and has been replaced by metal and plastic containers that now serve its erstwhile purpose.

Bamboo and Cane Utilisation in Traditional Craft

Some traditional basketry items still have a local market and usage but overall, there is a huge drop in the use of traditional craft products, along with the inevitable lifestyle changes. As a livelihood option, which it never was, basketry is not found to be sustainable. But having said that, it is extremely important to underline the immense value of this fine skill that makes this craft tradition so precious. The craft can be adapted to the new context with creative and minor design inputs to ensure the preservation of basketry skills that are now highly endangered and rarely found except amongst a few village elders. The main concern today is the loss of undocumented traditional knowledge and high-quality practical hand skills related to the making of these products which, eventually, if lost will become wholly extinct. Incorporating craft-skill learning in local school curriculum is vital to the preservation of this languishing craft tradition.

5 Vulnerability and Hope

Figure 5.0 Bird's eye view of Ziro Valley, 2021.
Source: Dr Tage Kanno, Hapoli.

DOI: 10.4324/9781003305538-5

The Apatani Way of Life

This is the story of how a material culture was shaped around bamboo and cane resources and sustained by a strong community spirit and spirituality. This elusive spiritual essence was one that transcended the human world, immersed itself in nature and its mystical life force and thereby sustained an unbroken ethos of conservation for untold centuries. The crafting of bamboo, cane and land use defined this way of life and created an architectural language of its own. This defined the settlement and its structural design. The landscape of the built and the natural came together harmoniously and deliberately.

This study delves into fascinating indigenous knowledge systems built around field and forest and takes a deep look into a now well-proven, uniquely sustainable way of life that continues to evolve in the current global context. Like all indigenous tribal knowledge, this degree of evolution relies on centuries of accumulated intimacy with the natural environment. It is about how bamboo houses can be built in three days, and how fish is cultivated in a rice field. It teaches you how to plant and look after a bamboo garden and how a single river can feed an entire community for millennia; how the human spirit can engineer an agricultural landscape out of a swamp and how peaceful coexistence can withstand the trials of time. It records for time to come, a stunning array of aesthetically handcrafted bamboo-cane basketry and a famous textile-weaving tradition that is now in an endangered state. It is about an extraordinary legacy of a people who have never lost their deep connect with nature. This includes lovely bamboo gardens, well-irrigated rice fields, sacred forests and a forestry tradition that protects the natural forest cover. The landscape remains undisturbed because the community recognises the value and importance of these natural gifts and the fact that they are irreplaceable once lost. This is further anchored by a strong community spirit that prides itself on the beauty of its harmoniously engineered landscape and a nature-based spiritual faith.

As a stand-alone example of a deeply imbibed conservationist outlook, it is a beacon of inspiration to all humankind. However, the constant pull and push of the technical age has a disconcerting way of unhinging all naturally sustainable indigenous lifestyles. This lurking vulnerability is very real and can definitely pose a threat to the Apatani way of life as well. One cannot gloat over this lifestyle's famed resilience and erase that possibility entirely. On the other hand, there is hope that, amidst battering changes, all is not lost. What lies behind both this hope and vulnerability, however, needs conscious and objective examination and a conviction that its relevance is undying.

Adapting to Change

It is important to dissect what actually defines the huge global relevance of the Apatani way of life. A deep understanding of what has inspired and nurtured this ethos becomes necessary if one must find hope for its continued evolution in an increasingly destructive and wayward global landscape. The Apatani core dwells in an environmentally and socially conscious attitude to life. This sensibility and sense of responsibility are by-products of a spiritual and ethical mindset that has evolved over centuries of close contact with the natural environment. That sensibility has been furthered by the historical circumstances of limitation of resources and land, as mentioned earlier. In stark contrast to the current, all-pervasive exploitative global atmosphere that surrounds this community, it has amazingly retained its well-preserved forest cover; under strict observance of social and communal norms that continue to find unusual respect, even amongst their youth. The Apatani have been able to preserve their longstanding ecology, governed by age-old customary laws – uncoded, but well-applied – that control and regulate land and forest use and are further enforced by the *Bulyañ* or traditional Village Council. The rule of the *Bulyañ* at village level has been strongly effective in contributing to an unshaken ecosystem, though its current relationship with competing government officials causes tension in some areas of governance. The role of the Council is now limited to specific village-level governance and so its effectiveness in enforcing law and order has diminished. This fact is felt, and often articulated, by the community especially during discussions on changes in the socio-political structure.

The traditional well-conceived land-use system has similarly endured a fully altered context of extended settlements and new forms of governance and politics, along with the well-recognised and highly consumerist external influences. Even though the land under agriculture has doubled in the past sixty years, it has never been at the expense of the community-owned forest but rather, atypically, based on the conversion of foothill wastelands that adjoin the forested hills. The signature rice and fish dual cultivation practice of the Apatani continues to retain its high yields and fertility and is the pride of the community. The ancient and highly efficient irrigation channels continue to water the well-kept fields, although a fee is now levied for the maintenance that was once done by cooperative effort. Similarly, the bamboo and timber plantations have withstood time and its effect, even though there is a considerably diminished use of bamboo in both construction and fencing.

Commercially oriented new cropping patterns that have introduced crops like kiwi, cardamom and pepper have entered these plantation plots and are the more noticeable changes in the agricultural system. What is new in this dimension is an extension and understanding of the commercial aspects of agriculture like floriculture (orchid farming) and horticultural practices. These do not seem to disrupt but rather find space within the traditional land-use system. Interestingly, and very fortunately, the consumerism that is associated with business and finance do not seem to be able to unravel this consistent eco-culture, even though they affect these minor changes within it. Overall, both field and forest remain mostly untouched amidst the (in contrast) transformed built settlement area. All this has been made possible by the stubborn resilience of the Apatani social ethos that holds fast onto its strong unity and self-sufficient community spirit.

A Changed Architectural Landscape

Today, despite the usual government interventions and the much-improved communication, the real indications of change that are apparent in the last twenty or so years are largely in the settlement zone and built structure (see Figure 5.1). All in all, it is the built environment that has succumbed to change and is largely deviant from its earlier *avatar* in both façade and form. Newly built and differently designed houses have replaced the typical bamboo-based, one-hall plan shelters of the vernacular architecture. To disperse the close clustering that suggested an omnipresent fire hazard, buildings are now dispersed over a larger built-up area. Some newly constructed houses have now begun to sit inside grazing pastures, kitchen gardens and, although very minimally, even in some agricultural fields. Though cement and all the less attractive materials of the modern era have seeped into the architectural construct of the Ziro Valley; they still must compete with both bamboo and timber, which continue to be used in conjunction with these, albeit differently.

Although the use of bamboo in building construction has lessened in the last ten years; it continues to find a place in the currently modified architectural design style. Its usage, however, has been modified to increase the

Figure 5.1 Collage of the current built and natural landscape (Collage of three images).
Source: Dr Tage Kanno, Hapoli.

durability and life of the built structure and adapt it to the contemporary lifestyle. Thus, bamboo is mostly retained in the interior elements and roofing structure of the new architectural designs that now pervade the built environment. Inset wall mat panels within the timber or RCC frame and inner partition walls are still constructed mostly with bamboo.

This marked retention of bamboo elements amidst the hugely altered built environment that continues to be a part of the architectural language and building ethos, is a favour that comes from centuries of knowledge, familiarity, and identity. Of course, the easy availability and natural properties of this resource, and a marked recognition of its contribution to the building of their community ethos, have contributed to its social significance. And its ritual use in festivals and ceremonies continues to remain as strong as ever.

The adoption of corrugated galvanised and laminated sheet metal roofing in place of the bamboo culm planks – or half-bamboo tiled roofs and reinforced concrete raised pedestals instead of the traditional bamboo or timber stilts – was a major visible alteration that had come in by the late nineties. Nowadays these are the norm and, with them the invariable brick or concrete foundations and RCC frames that support an infill timber or bamboo walling are other preferred building styles that have replaced the all-bamboo vernacular house. Overall, a composite structure is seemingly preferred, and it most often stays with the woven bamboo *tarza* (see Glossary) or timber plank walling along with basic structural bamboo or timber roof truss elements.

As a whole, the centuries-old scenic natural landscape remains largely unruffled (Figure 5.1). During his 1962 revisit to the area after a 17-year-long gap, the famous anthropologist Christoph von Furer-Haimendorf expressed the surprised assertion of this same fact, reinforcing the fact of sustainability of the Apatani way of life.

A very interesting and pertinent excerpt from Von Furer-Haimendorf's book, *The Apa Tanis and Their Neighbours*, is a comment on change or the lack of it, shared below.

> *Walking through the streets of an Apa Tani village or sitting around open fires of one of the dark, cosy houses one notices hardly any change compared to the days when the valley had only the most tenuous contacts with the outside world. And this lack of change is not confined to the material aspects of the Apa Tani's life. Their discipline and sense of social responsibility [have] enabled them to maintain the autonomy of villages almost unimpaired and a sympathetic administration anxious to interfere as little as possible in the tribe's internal affairs, has welcomed this self-sufficiency of the Apa Tanis.*
>
> *The restraint of the Apa Tanis in acquiring novel consumer goods is remarkable. Even in the house of men who through trade or wage labour have earned considerable sums of money, one sees few articles of foreign*

manufacture. There are usually some iron cooking pots in place of the old earthern pots, metal buckets are gradually replacing gourd vessels and tin boxes with padlocks are used instead of store-baskets. But otherwise, the Apa Tanis have been slow in developing a taste for foreign goods and their living habits and diet have remained almost unchanged. The conservatism of the Apa Tani women in the manner of dress differs noticeably from the Dafla (Nyishi) women, many of whom wear machine made textiles including sometimes even readymade dresses in western style.

(Christoph von Furer-Haimendorf, 1962, *The Apa Tanis and their Neighbours; A Primitive Society of the Eastern Himalayas*. London: Routledge and Kegan Paul)

My own witnessed account concurs hugely with Von Furer-Haimendorf's 1962 narrative, as what I experienced and documented in 1999, over thirty-five years after him, largely resembled his account! Interestingly, now more than twenty years later, changes in some aspects of the social landscape though obviously more pronounced, still retain the sense of restraint and consciousness that he flags so accurately in his book.

Evolution, Not Dissolution with Change: A Local Perspective

- *What change do you see in the settlement or built environment in the past 20 years? What in your view are the reasons behind these changes?*
- *Are you hopeful of an evolution rather than a dissolution of the famous ethos of the Valley? Do you think it has weakened in the past decade or strengthened because of global recognition?*
- *What do you think is the greatest threat to the Apatani cultural landscape and why?*

Twenty years later, in 2021, I was in conversation with Dr Tage Kanno of Ziro, to whom I put all these difficult questions. He tackled each question beautifully and briefly gave a clear and concise perspective on changes in the Valley and on the current scenario. He is a medical doctor and a long-time member of Future Generations, an international US-based NGO. A Facebook group named 'Ziro Calling You' is one of his initiatives, along with some others. This forum is a useful and interesting link that shares current photographs of the Valley with interested viewers across the globe. 'Ziro Calling You' works as a kind of notice board that enables non-residents to keep in touch with local events and changes in the Valley and with locals or visitors to contribute their photos to allow the same.

Dr Kanno has a keen sense of community and recognition of his valuable heritage. He has two extremely important initiatives to his credit. The first is the nomination of Ziro Valley as a UNESCO World Heritage Site to attract global recognition of the Apatani way of life. It has now been included in

UNESCO's Tentative List as the Apatani Cultural Landscape (https://whc.unesco.org/en/tentativelists/5893/). The second initiative was to recommend the use of a modified Roman script for writing in the Apatani language, something that has been partially accepted by the community (http://language.apatani.org/). Dr Kanno is also mindfully cognisant of the kinds and effects of change that shadow the cultural landscape of the Ziro Valley. I had reconnected with him after a long gap of twenty years to once again relive my wonderful memories of the Valley in 1999, when I first met him there. The discussion with him was for me to get a sense of what the Valley is like today for those who live there and who have also seen what I saw and documented so many years before.

To explore and highlight local perceptions of far-reaching changes beyond my own is, I believe, particularly important for this kind of book to address at this point in time. It is a time of the in-between, where the past and present still coexist and continue to evolve in a continuum. Within this evolution, there are the ever-lurking fears of dissolution and along with them also the hope of continued growth. Both must necessarily coexist in relation to such a resolute and deeply cherished heritage.

Dr Kanno says that the Facebook group was created to enable a bond with the Valley and its now internationally famous ethos. He believes that, although change is a certainty and a permanent reality that cannot be fully controlled, the Valley's fundamental nature has been one of evolution rather than dissolution of the Apatani culture. Like me, he has the strong conviction that the Apatani way of life shows the way and has something special to offer to the world. It can be an inspiration for the rest of India, the world at large and, more importantly, also for the Apatani youth. It has proved itself to be a harmonious and sustainable lifestyle and, nowadays, this word is understood and used everywhere, Dr Kanno says. Any trend cannot be stopped consciously, but we can now talk about sustainable development and be heard clearly. There is an understanding of the need for going back to the old way, at least in spirit, and that this realisation is coming home to the younger generation – which is now very visible in Ziro.

I wrote about this after my interview with Dr Kanno.

> *The undoubted shifts in context, lifestyle habits and building trends in the past 10 to 15 years have not eroded the cultural ethos and value system of the Apatani community. Things have changed but the base remains intact. Everything is inter-related and nothing ever remains the same. Many members of the community have adopted Christianity which has slightly diminished the adherence to the nature-based rituals based on the ancient faith and belief system (Donyi Polo). The number of shaman priests or nyibu is gradually decreasing as the new generation is unwilling to embrace this role, so the ritual ceremonies and other aspects that were conducted solely by them are consequently disappearing gradually and visibly.*

There is both material and design change in the built environment. The obvious reasons for these changes lie in an altered lifestyle that brings about its own demands that accompany a younger mindset. The vernacular house type had a verandah and single roomed plan which has become unsuitable in the altered context of contemporary perceived needs. Separate rooms and privacy are now a necessity like a demarcated kitchen, a living room and different bedrooms or even quiet study areas for school going students. The emphasis on modern systems of education is an undeniable felt need and must be accommodated in the new architectural language. It is the spiral effect of many such interdependent factors that fuels changes of this nature and each one knocks down the other in a domino effect of sorts. Additionally, the availability of new construction materials has been made possible with the much-improved connectivity and easy availability of water and electricity supply in all villages in the past 15 years. These typical industrial materials like cement, brick and iron sheets have now become the obvious preference primarily for their comparatively high durability and minimal maintenance qualities. The new need for minimal maintenance is also something that is a direct outcome of a changed lifestyle that dictates a more individualistic philosophy and the lessening of manual labour. One of the most visible effects of the modern era is the loss of the tradition of community building which has all but vanished now. This is the definitely the biggest change that one can see in the traditional societal cooperative structure. However, mindful of the far-reaching consequences of this loss of community bonding, the community has now created a house building fund. This fund sustains the clan group support system that otherwise ensured the coming together of all the clan members to help construct the traditional bamboo-based house. As a new version of the old system, it provides monetary support to any clan member wanting to either build or renovate his or her house. All clan members contribute equally and regularly to this building fund and can apply to it in times of need. There is a marked change in the vernacular settlement pattern as some new houses are even being constructed in the once untouched rice fields due to the increasing scarcity of land and a growing population.

The threat to the Apatani Cultural Landscape has become more pronounced in the last 10 odd years. This is due to government interventions in the name of 'development' which have encroached into the domain of land use in some ways. There has been a deliberate attempt to resettle people into newly established settlements as in Mani Polyang, Dilo Polyang and Hapoli to name a few. However, it has to be said that a few people themselves have on their own also opted for new building trends and sites.

A considerable number of young people tend to leave the Valley to expand their horizons in terms of learning and experience, but they also always return to contribute, never abandoning their ethnic roots. The ones that leave are seen to be more conscious of the value of their culture. They always want to come back home. They have not lost their strong emotive connect to the Valley. They make new houses in the villages and do not migrate to live outside permanently. Timber and concrete houses with galvanized iron sheet roofing are now the latest norm and bamboo is now mostly used in interior work. The bamboo groves continue to be maintained as before even though bamboo roofing has been wholly abandoned due to it being a fire hazard and having a short lifespan.

Bamboo will be used for a long time to come as we have a special cultural relationship with it. *Evolution is happening, not dissolution in spite of all these changes. We can never entirely control the kind of changes that happen and must accept them to some extent. Global recognition has not affected our people. Some people did not want this Cultural Heritage label as they wrongly interpret it as a hindrance to development of the area. They think that our Valley will become like an untouched museum piece and all forms of development will be stalled because of the heritage label. This is the kind of attitude too that we are sometimes faced with and have to deal with at times too. Nothing has really changed or happened on the ground as yet even with all this global recognition of sorts.*

(*Excerpts from the author's interview with Dr Tage Kanno, Hapoli, 2021*)

Recognising Vulnerability and Hope

Nothing can be forever, and contexts must keep changing as is the nature of existence in this world. It is understood that change happens either in the form of gradual growth or suddenly with sometimes disastrous effect. Today even as the world is in crisis mode around increasingly evident climate change signals and serious environmental damage, the Apatani Valley might still just weather the millennial storm. Its ethical resilience in the wake of the dominating influences of the millennial age is highly unusual. Though the aforementioned current indications point at endangering the people of the Valley's long-sustained harmonious relationship with the natural environs, a mindfulness and understanding of their invaluable contribution to the health of the planet could turn the tide. The recent international spotlight on the community, related to the currently critical issues of climate change, would possibly help direct available resources towards further securing its future to some extent. This is not to suggest stagnation, but moving onwards with consciousness and value for a spirit and system that has sustained the green legacy of a people.

While field and forest still remain largely untouched as of this writing (2021), the settlements are naturally transformed and unrecognisable from twenty years ago. This is natural and bound to happen. There are increased efforts towards the building of educational and health facilities. In 1999, too, forest expeditions had become a rarity and hunting and cane and wild berry collection had become occasional. Nowadays, the rarity of these forays is further pronounced, as the use of air guns has also been recently stopped by the state government to dissuade game hunting of any kind. The forest is now the haven of trekkers and hikers! Ziro is now quite the eco-tourist destination and has all the paraphernalia that goes with that, like burgeoning campsites and village homestays that promote eco-tourism. This has been further accentuated by the hugely transformed connectivity of the state in general. Without judgement, as an objective statement of fact, this could obviously go both ways.

There is fundamentally a limited area for a growing populace that strains the seams of the settlement area, so that there is an ever-lurking concern that, in the years to come, the seams will burst and settlement will occupy larger parts of the agricultural area. In the end, what matters will be a complete awareness of the possible dangers that threaten the eco-culture of the Valley. This is the only way out and, coupled with it is a recognition of what is to be valued more than anything else.

In so many of the other cultures of the Indian subcontinent, a sustainable hands-on lifestyle is shed as easily as snake's skin, whereas non-contextual ritual is clung to mindlessly for the most part. Today, everyone can and should learn a lesson or two from this admirable community. I believe, as a witness to such steadfastness, that it is the strong community spirit and deeper ethical grounding that reinforces this choice and differentiates them from the vast majority of humanity on this planet today. In this grounding, lies the eternal hope and faith that the longstanding Apatani ethos is strong enough to continue its dogged resistance as it has for so long. Though there is undoubtedly an increasing recognition of their vulnerability to external influences and globalisation; there is also a defiant spirit that believes it can evolve within this challenge and not be bulldozed.

References

Blackburn, Stuart. 2010. *The Sun Rises: A Shaman's Chant, Ritual Exchange and Fertility in the Apatani Valley*. Leiden, Netherlands: Brill.

Rechlin, Michael. A. and Ritu Varuni. 2000. A Passion for Pine: Forest Conservation Practices of the Apatani People of Arunachal Pradesh. *Asian Studies Journal,* vol. 26, no. 1 and 2, pp. 19–24, University of Michigan.

Varuni, Ritu. 1999. *Identification and utilisation of bamboo and cane resources and related land use patterns in the Apatani plateau; An analytical study and documentation*. Unpublished personal field research project report, funded by G.B Pant Institute of Himalayan Development and Environment, N.E Unit, Itanagar.

Von Furer-Haimendorf, Christoph. 2010. *The Apa Tanis and their Neighbours; A Primitive Society of the Eastern Himalayas*. London: Routledge and Kegan Paul, 1962.

NOTE: The present book is based on the author's field research project and study analysis done in 1999 and thereafter. It has been updated to the current context. The author lived in Ziro Valley for four months in 1998–1999 to undertake this project and for six years in Arunachal Pradesh as a whole. She also studied Apatani vernacular architecture in depth over the 1990s and recorded changes therein from 1993 to 1999. In 2021, she conducted interviews with people in the Ziro Valley to acquire an overview of perceived changes that have taken place in the built environment.

Glossary

Apatani Terms

aji: Wet rice field plot
bachiñ: Yellow-green sour-tasting wild berry
balu: Kitchen garden
bije: Bamboo variety (phyllostachys bambusoides) or bamboo garden
Bulyañ: Traditional Apatani Village Council
byago: An open verandah at the entrance of the vernacular Apatani house
byaku: Onion
byayuñ: Potato
byayuñ: Tomato
byopa: Traditional finely woven cane helmet-like hat or headgear worn by men
embiñ: Husked rice
emo, mipya, pyapiñ: Indigenous rice varieties
giyañ hamañ: Mustard variety that grows twelve months of the year.
hi': Dried bamboo shoot
hikhu': Fermented bamboo shoot
hula': Grape-like fruit with a sweet-sour flavour
ilyo: Or *Dao* (Assamese), traditional multi-tool machete with wooden handle
kung: Nut of the oak variety
kwdi: Land
kwra: Oak tree
le'ra: Traditional woven cane backpack used mostly on hunting expeditions
lya'pyo: Millet plot
mampo yamyo: Flattened heat-bent bamboo culm ridging in the vernacular Apatani house
midiñ: Rice seedling nursery
mipya: Early-ripening variety of millet
miyu/tani: Human being
more: A community- or clan-owned forest
muku: Tobacco
mwge': Bamboo tongs

myopu: Flooring of flattened bamboo in the vernacular Apatani house
nyatu: Rest point in the forest, marked by a single planted blue pine and a timber bench
Nyibu: Shaman-like priests of indigenous Donyi-Polo faith of Abotani tribal group
o: Rice beer
obyo peruñ: French beans
pecha: Wild apple tree
piñta: Marrow
potuñ peruñ: Soya bean
pwsa: Blue pine
pwta': Pear
re'ke: Bamboo loft in the vernacular Apatani house
salyo: Fruit peel of the Michelia tree (Oak species), which is ground into a chutney
sampvr: Bittersweet wild blackberry variety
sañkhe': Nut of the oak variety
sansuñ/salw: Timber plantation
sarse: Late-ripening variety of millet
semo/señbo: Red berries of the forest
subu: *Mithun* or Indian bison (Bos fontalis)
sudu: Or *chunga* (Assamese), bamboo container measure used for the sale of bamboo shoot
ta'miñ: Sweet and somewhat salty tasting wild fruit
taku: Cucumber
takuñ: Peach
tanyi': Maize
tape: Pumpkin
tapyo: Traditional Apatani black salt-like substance from extract of a leafy forest fern
uko: Back verandah of the vernacular Apatani house
yaso: Cane
yorlu: Vegetable maize plot

Other Terms

Abotani group of tribes: Tribal communities from Arunachal Pradesh, who share a common lineage and ancestor belief. They are said to be the descendants of Abotani, the primal ancestor, and are also referred to as the Tani tribes, comprising the Adi, Galo, Tagin, Nyishi, Apatani and Mishing (in Assam) communities. These origination stories are a major part of the oral tradition of these communities.
Bije (Apatani origin): Phyllostachys bambusoides; a monopodial bamboo species propagated by the Apatani people.

Dao: Multipurpose machete used for all manner of bamboo and wood work and which is the principal traditional hand tool of the Tanii group of tribes in Arunachal Pradesh. It is called *ilyo* in Apatani.

Jhuming (Bengali origin): A slash-and-burn way of farming or shifting agriculture practised by many communities in Eastern India, where a land plot is cultivated for some years and then left fallow for a number of years to recover while a new one is cleared for farming.

Jhumlands Regulation of 1947: This act, passed in 1947 by the Government of India, gives sanctity to the customary right to *jhum* land or to practise slash-and-burn agriculture in favour of a village or community while providing the legal basis for land administration. Within this regulation, all customary laws and traditions are deemed to be respected, and village authority prevails over all else.

Machaang (Assamese origin): Open raised platform made of bamboo, usually attached to the house.

Tarza (Assamese origin): Flattened bamboo culm planks woven together and used for walling and flooring.

Index

For Product Safety Concerns and Information please contact our EU
representative GPSR@taylorandfrancis.com
Taylor & Francis Verlag GmbH, Kaufingerstraße 24, 80331 München, Germany

www.ingramcontent.com/pod-product-compliance
Ingram Content Group UK Ltd.
Pitfield, Milton Keynes, MK11 3LW, UK
UKHW020208200926
475702UK00004B/671
* 9 7 8 1 0 3 2 2 4 7 1 3 7 *